25x
UPPER-ELEMENTARY: STEM-ENGINEERING

Workbook #3 of 4 in the STEM Series
for Upper Elementary Students

version 1.0 - 11-15-2024
Produced by 3andB Inc. | 3andB.com
For more information: info@3andB.com

Welcome to Your STEM Engineering Adventure!

Dear Student,

Get ready to explore the amazing world of engineering! This workbook is your guide to discovering how things work - from simple machines to airplanes. You'll learn about engineering fundamentals, mechanical engineering, and modern engineering, and so much more through fun activities and readings.

How to Use This Workbook

1. Read Carefully
- Take your time reading each chapter's introduction
- Look for bold words - these are important science terms to remember
- Ask questions if something isn't clear

2. Complete All Activities
- Fill in all blanks and answer all questions
- Use complete sentences for written answers
- Keep your work neat and organized

3. Have Fun with Word Games
- Each chapter has word scrambles and puzzles
- These games help you remember important science terms
- Try to solve them on your own before asking for help

4. Check Your Work
- Use the Answer Key online to check your answers
- If you make a mistake, try to understand why
- Remember: making mistakes is part of learning!

5. Keep Track
- Write your name on your workbook
- Date your work as you complete each section
- Keep all pages clean and in order

Important Tips:
- Always have a pencil and eraser ready
- Take breaks between chapters
- Discuss what you learn with classmates and family
- Draw pictures to help you understand concepts
- Most importantly - be curious and enjoy learning!

https://3andb.com/answer-keys/

Remember: Engineering is all about discovering new things and understanding our world better. Don't be afraid to make mistakes or ask questions. That's how all great engineers learn!

Let's begin our engineering adventure together!

25x: UPPER ELEMENTARY: ENGINEERING

Engineering is the creative process of designing and building things, like bridges and robots, to solve problems and make our lives better!

Engineers are amazing problem solvers who use their creativity to make our world better! An **ENGINEER** starts by identifying a problem that needs a **SOLUTION**. They create a **DESIGN** and draw a detailed **BLUEPRINT** to show their ideas. This is all part of the **PROCESS** of engineering. Before building the final product, engineers often create a **PROTOTYPE** to test their ideas. This first model helps them spot any problems early. **MECHANICAL** systems, like robots or machines, need extra careful testing to make sure all moving parts work correctly.

TECHNOLOGY plays a huge role in modern engineering. Today's engineers use computers and special tools to create amazing **INNOVATIONS** that help people. Whether they're designing a tall **STRUCTURE** like a skyscraper, inventing a new medical device, or creating a faster computer, engineers are always thinking of new ways to improve things.

Engineering is all about being creative, solving problems, and never giving up when things get challenging!

WORD SCRAMBLE

#	Scramble	Answer
1	OSCRPSE	
2	NOINONIVAT	
3	EINDGS	
4	ENEREIGN	
5	AALECNMHIC	
6	OUSTNLOI	
7	ECOTNYHOGL	
8	RILBPEUNT	
9	SEUUTRRTC	
10	YORTOEPPT	

Technology and engineering work together to create amazing tools and machines that help us solve problems and make everyday tasks easier!

I. Key Definitions

Write the definitions for these important terms:

1. Engineer: _____

2. Prototype: _____

3. Innovation: _____

II. Fill in the Blank

Word Bank: BLUEPRINT   MECHANICAL   PROCESS   SOLUTION   TECHNOLOGY

1. An engineer starts by finding a problem that needs a _____.

2. Engineers create a detailed _____ to show their ideas.

3. Engineering follows a _____ to create new things.

4. _____ systems, like robots, need careful testing.

5. _____ helps modern engineers do their work.

III. True or False

Mark each statement True (T) or False (F):

1. _____ Engineers use creativity to solve problems.

2. _____ Engineers only work on buildings.

3. _____ Prototypes help engineers find problems early.

4. _____ Engineers give up when things get challenging.

IV. Matching

Match the items in Column A with their related items in Column B:

1. _____ Engineers   A. Tests ideas

2. _____ Prototype   B. Problem solvers

3. _____ Computers   C. Used for testing

4. _____ Blueprint   D. Modern tools

V. Reflection

Why do you think engineers are important for making our world better? (Write 2-3 sentences)

_____

_____

_____

What is Engineering?

Across

2. This is something that is built, like a bridge or building

7. A detailed technical drawing showing how to build something

9. Tools, machines, and inventions that make work easier

10. A person who designs and builds things to solve problems

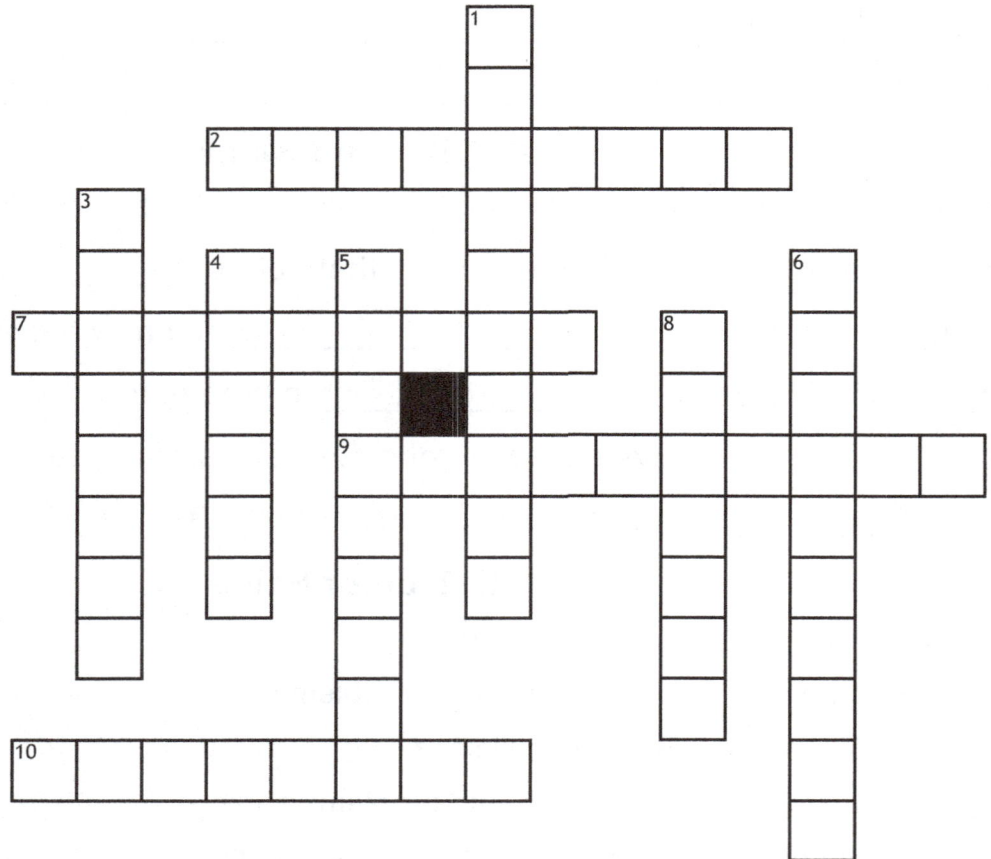

Down

1. Related to machines and moving parts

3. The answer to a problem that needs to be fixed

4. A detailed plan for making something

5. The first model of something that can be tested

6. A new idea or way of doing something

8. A series of steps to complete a task

What is Engineering?

```
P R O C E S S G S V Y T J J U R F
O A O N M B M J R O N S S F U B E
Y B B S Z E J E Z A L O Q W I C S
T V F T T Q A Z C D F U S E G S D
N T C E O R S D F H E D T U P A O
Z E U C X G U Q K U A S I I T W X
P N H H G X L C R X G N I J O J P
R G E N J V L O T W N F I G E N S
O I J O H R H B W U M Q N C N W C
T N L L E K B L V C R L N K A C U
O E C O R M K U M G L E O H F L R
T E J G T M S E A F I H V I N J Q
Y R L Y I A L P U M V H A Z S E W
P D P Y G F Z R F N F J T U K H W
E J O H U B E I P G A V I J J O H
M N D G J Z A N D R P K O M J A T
G K W M R W M T T S A O N Q P S M
```

STRUCTURE   INNOVATION   PROCESS

MECHANICAL   TECHNOLOGY   SOLUTION

ENGINEER   PROTOTYPE   BLUEPRINT

DESIGN

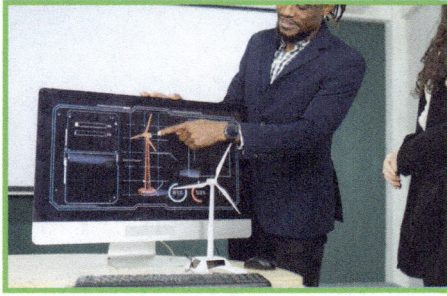

The engineering design process is a series of steps that engineers use to come up with ideas, build prototypes, test them, and improve their designs to solve problems!

When engineers work to make our world better, they follow the Engineering Design Process. It all starts when they spot a **PROBLEM** that needs fixing. They **ANALYZE** the situation carefully to understand what's going wrong. Then they **BRAINSTORM** lots of creative ideas - the more, the better!

Engineers know it's important to **COLLABORATE** with others because teamwork leads to better ideas. Once they choose their best idea, they create a **DESIGN** that shows exactly how their invention will work. The next exciting step is building a **PROTOTYPE** to test their idea in the real world.

They carefully **EVALUATE** how well their prototype works. Does it solve the problem? Is it safe? Can it be made better? Engineers always look for ways to **IMPROVE** their work until they find the perfect **SOLUTION**. That's what makes a great **ENGINEER** - they never give up until the job is done right!

WORD SCRAMBLE

#	Scramble	
1	TOOSINUL	
2	ABRTMOISNR	
3	NGEERINE	
4	ULVAAEET	
5	EPMIVOR	
6	IENGDS	
7	OEPTTYRPO	
8	MORELPB	
9	CLBEOLAATOR	
10	ALZAEYN	

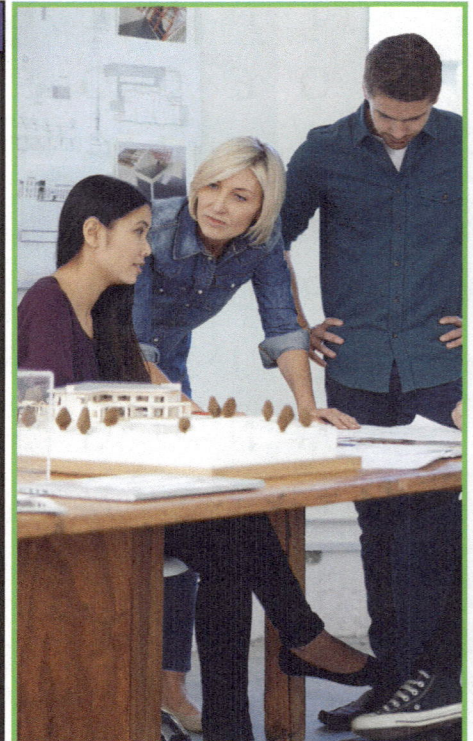

Prototyping is making a simple model of an idea or invention so that we can see how it works and figure out how to make it better!

I. Key Definitions

Write the definitions for these important terms:

1. Problem: _____

2. Prototype: _____

3. Engineer: _____

II. Fill in the Blank

Word Bank: ANALYZE   BRAINSTORM   COLLABORATE   EVALUATE   IMPROVE

1. Engineers _____ the situation to understand what's wrong.

2. Engineers _____ many creative ideas to solve problems.

3. Engineers _____ with others because teamwork makes better ideas.

4. Engineers _____ how well their prototype works.

5. Engineers always look for ways to _____ their work.

III. True or False

Mark each statement True (T) or False (F):

1. _____ Engineers work alone to solve problems.

2. _____ Engineers test their ideas with prototypes.

3. _____ Engineers give up when things get difficult.

4. _____ Engineers look for ways to make their work better.

IV. Matching

Match the steps of the Engineering Design Process:

1. _____ First Step   A. Test the prototype

2. _____ Second Step   B. Find a problem

3. _____ Third Step   C. Analyze the situation

4. _____ Fourth Step   D. Brainstorm ideas

V. Reflection

Why do you think it's important for engineers to test their ideas before making the final product? (Write 2-3 sentences)

_____

_____

_____

Engineering Design Process

Across

2. To make changes that make something better

6. A first model used to test an idea

8. To study something carefully to understand it better

9. To think of many different possible solutions

10. A person who designs and builds things to solve problems

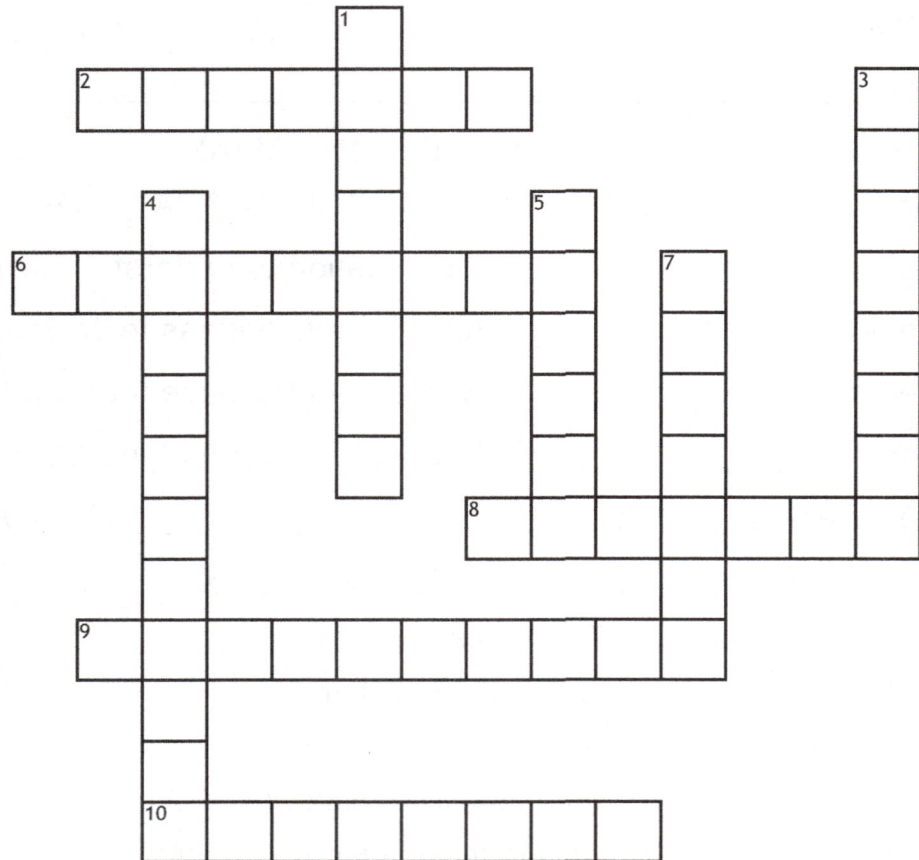

Down

1. The answer or fix to a problem

3. To carefully check how well something works

4. To work together with others to solve a problem

5. A detailed plan showing how to make something

7. A situation that needs to be solved or fixed

Engineering Design Process

```
T P D Q N W S B V S E X Z Y I I H
C R D W L X A I R K Y O P E V O A
D O D D W R R Y I A E N K G K D Q
I T G I K T Q M J D I M G P Y K L
B O E O R J D H T R R N B W U S Q
E T A C K J K S Z O E A S R R F U
H Y S O Q E N G I N E E R T N B U
S P D L W R S A B W E Y G L O J G
D E O L N X S O L U T I O N G R E
V U B A P R O B L E M O A V M Q M
V M B B I K Q X D P A K N B O S S
F X J O D M L Y S E O N M O G I X
Z B I R L R P M M B S Q A W X J V
X V R A S T O R R K A I Z L P C V
K X A T B F V H O L B G G V Y E N
G B X E R X Z L C V V X N N C Z S
I Z R E V A L U A T E J V T N O E
```

ANALYZE   ENGINEER   COLLABORATE

SOLUTION   IMPROVE   EVALUATE

PROTOTYPE   DESIGN   BRAINSTORM

PROBLEM

Problem-solving methods are different ways we can think through challenges and find solutions, like breaking the problem into smaller parts or trying out different ideas!

Good problem solvers know that having a **STRATEGY** is the key to success! When facing a tough challenge, they **BREAKDOWN** the problem into smaller pieces that are easier to handle. They also know the importance of **TEAMWORK** - getting help from others often leads to better solutions.

Before jumping in, smart problem solvers **RESEARCH** to learn what others have already discovered. They **ORGANIZE** their information carefully and **PREDICT** what might happen with different solutions. Then it's time to **EXPERIMENT** with different ideas!

As they work, they **OBSERVE** everything that happens and **DOCUMENT** their findings. This helps them remember what worked and what didn't. Finally, they **VERIFY** their solution to make sure it really fixed the problem.

These methods aren't just for scientists - we use them every day! Whether you're building a robot, solving a math puzzle, or figuring out why your bike makes a funny noise, these problem-solving skills will help you succeed.

	WORD SCRAMBLE	
1	SECHERRA	
2	EENRXMEIPT	
3	NGOIZERA	
4	TETAGYRS	
5	UEMTNODC	
6	RVIEYF	
7	NOERBAKDW	
8	CPTIDER	
9	ESBROVE	
10	WKROAMTE	

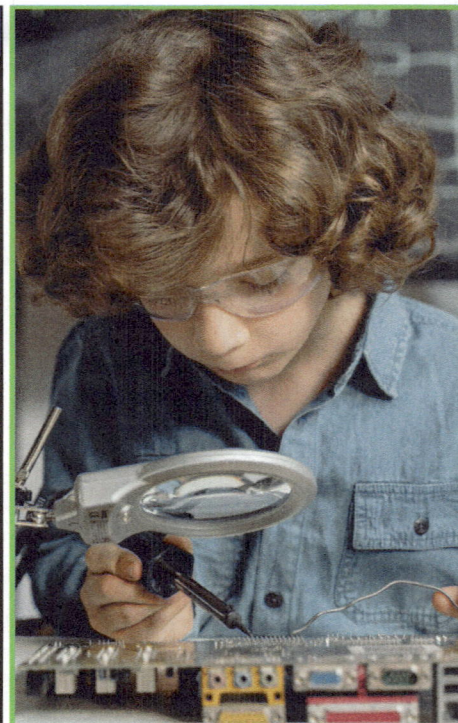

When we're solving a problem, observing means watching carefully, documenting means writing down what we find, and verifying means checking to make sure our solution works!

I. Key Definitions

Write the definitions for these important terms:

1. Strategy: _____

2. Research: _____

3. Teamwork: _____

II. Fill in the Blank

Word Bank: BREAKDOWN   DOCUMENT   OBSERVE   ORGANIZE   PREDICT

1. Good problem solvers _____ big problems into smaller pieces.

2. Problem solvers _____ their information carefully.

3. They _____ what might happen with different solutions.

4. As they work, they _____ everything that happens.

5. They _____ their findings to remember what worked.

III. True or False

Mark each statement True (T) or False (F):

1. _____ Problem-solving skills are only used by scientists.

2. _____ Working with others can lead to better solutions.

3. _____ You should try to solve the whole problem at once.

4. _____ It's important to verify if your solution really worked.

IV. Matching

Match the problem-solving steps with their purpose:

1. _____ Research   A. Keep track of what works

2. _____ Document   B. Make sure it's fixed

3. _____ Verify   C. Learn from others

4. _____ Observe   D. Watch what happens

V. Reflection

Think about a problem you solved recently. How did breaking it down into smaller pieces help you? (Write 2-3 sentences)

_____

_____

_____

Problem Solving Methods

Across

6. To split a big problem into smaller parts

7. To test different ways of doing something

9. To check if something is correct

10. To arrange information or items in a helpful way

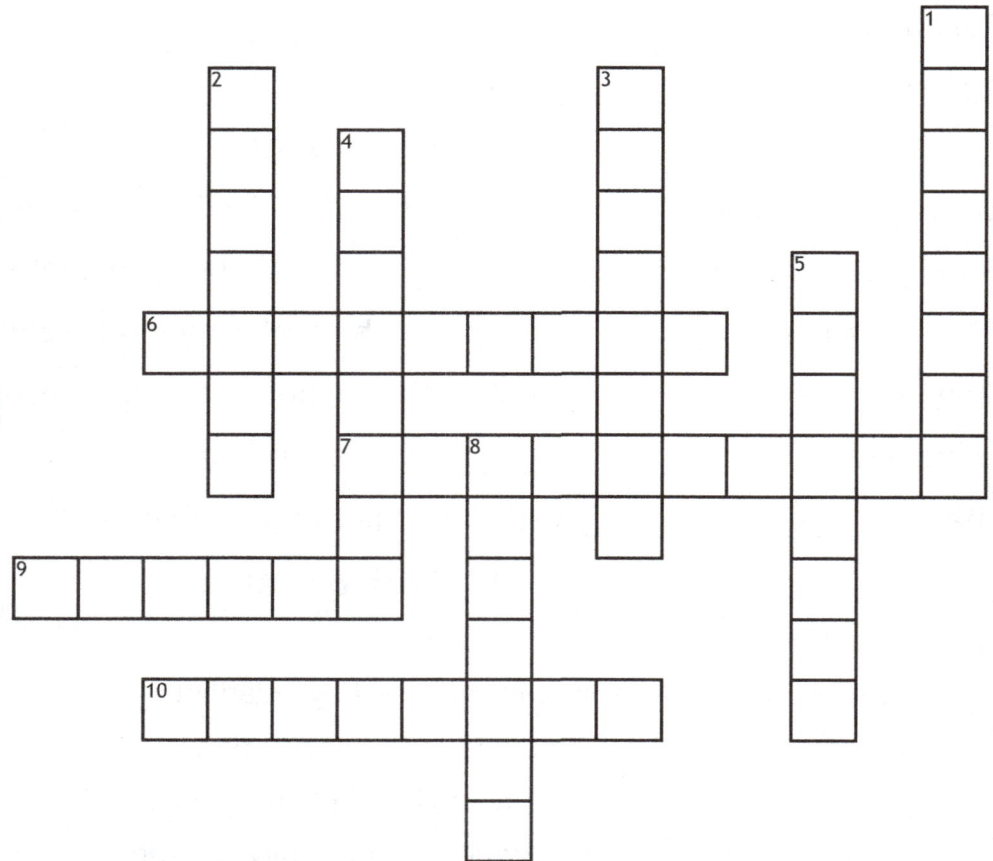

Down

1. To write down or record what you discover

2. To watch and notice details carefully

3. When people work together to solve problems

4. A careful plan to reach a goal

5. To find and study information about a topic

8. To make a guess about what might happen

Problem Solving Methods

```
L  L  L  E  D  E  Y  F  J  T  B  Y  R  J  I  M  D
N  D  Z  X  O  M  M  M  F  L  J  W  V  F  Q  M  B
F  K  I  P  C  E  L  P  R  E  D  I  C  T  F  J  W
H  Q  H  E  U  B  R  E  A  K  D  O  W  N  Y  N  C
Z  Z  R  R  M  P  B  F  T  O  J  E  L  A  D  S  H
I  H  C  I  E  O  G  R  Z  U  O  T  U  R  Q  T  Q
I  L  P  M  N  W  B  Z  R  K  K  N  D  W  F  R  N
B  H  B  E  T  O  O  V  E  R  I  F  Y  R  O  A  V
T  D  M  N  K  S  R  E  T  R  A  Z  W  L  V  T  I
C  W  H  T  Q  O  T  G  Y  E  E  Z  E  U  J  E  D
Q  F  B  V  E  B  Z  Z  A  H  A  S  T  M  H  G  M
B  I  O  I  K  S  S  N  Z  N  O  M  E  M  L  Y  L
B  H  E  O  S  E  Y  S  P  F  I  T  W  A  O  D  C
Y  R  O  T  Q  R  O  B  D  O  J  Z  S  O  R  K  O
U  Q  U  C  T  V  Y  Z  Q  E  A  X  E  A  R  C  E
P  Z  J  W  N  E  T  Q  E  P  F  V  H  I  L  K  H
Z  S  L  O  Z  N  E  G  D  O  F  L  V  K  H  Y  T
```

DOCUMENT	VERIFY	OBSERVE
PREDICT	ORGANIZE	EXPERIMENT
TEAMWORK	RESEARCH	BREAKDOWN
STRATEGY		

Measurement and tools help us figure out the size, length, or weight of things, making it easier to build and create accurately!

Scientists need to **MEASURE** things carefully to understand our world. That's why they use special **INSTRUMENTS** designed for different types of measurements. The **METRIC** system helps scientists around the world share their findings.

When you use a **RULER** to measure length or a **SCALE** to weigh something, you want your measurements to be trustworthy. **ACCURACY** tells us how close we are to the true measurement, while **PRECISION** means we can get the same result when we measure again and again. Scientists often **CALIBRATE** their tools to make sure they're working correctly.

A **THERMOMETER** helps us measure temperature accurately - important for everything from baking cookies to studying climate! When scientists measure **VOLUME**, they need to be extra careful because liquids can be tricky to measure. Whether you're doing a science experiment or building something cool, knowing how to use measuring **INSTRUMENTS** correctly will help

WORD SCRAMBLE

1	TTMISNRENU	
2	TMICER	
3	BAECLRIAT	
4	ERURL	
5	CSNIEIROP	
6	CCARYCAU	
7	EEMEOHTRMTR	
8	UVMLEO	
9	ALSCE	
10	REESMAU	

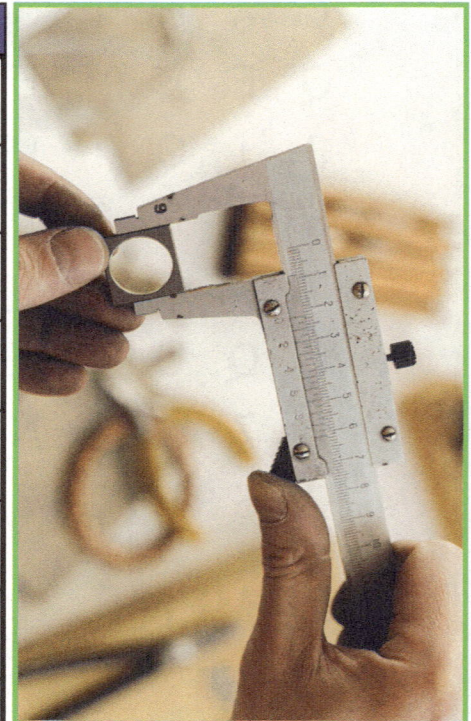

Calibration of instruments is the process of making sure tools, like scales and rulers, give accurate measurements so we can trust the results when we use them!

I. Key Definitions

Write the definitions for these important terms:

1. Accuracy: _____

2. Precision: _____

3. Calibrate: _____

II. Fill in the Blank

Word Bank: INSTRUMENTS   METRIC   RULER   THERMOMETER   VOLUME

1. Scientists use special _____ to take measurements.

2. The _____ system helps scientists share their findings.

3. A _____ is used to measure length.

4. A _____ measures temperature accurately.

5. Measuring _____ of liquids requires extra care.

III. True or False

Mark each statement True (T) or False (F):

1. _____ Scientists don't need to measure things carefully.

2. _____ A scale is used to weigh things.

3. _____ Measuring tools never need to be checked.

4. _____ Measurements should be trustworthy.

IV. Matching

Match the measuring tools with what they measure:

1. _____ Ruler   A. Weight

2. _____ Scale   B. Temperature

3. _____ Thermometer   C. Length

4. _____ Volume tool   D. Liquids

V. Reflection

Why is it important for scientists to take careful measurements? (Write 2-3 sentences)

_____

_____

_____

Measurement and Tools

Across

6. A tool that measures length or distance

7. A system of measurement used by scientists

9. A tool that measures how heavy something is

10. A tool used to take measurement

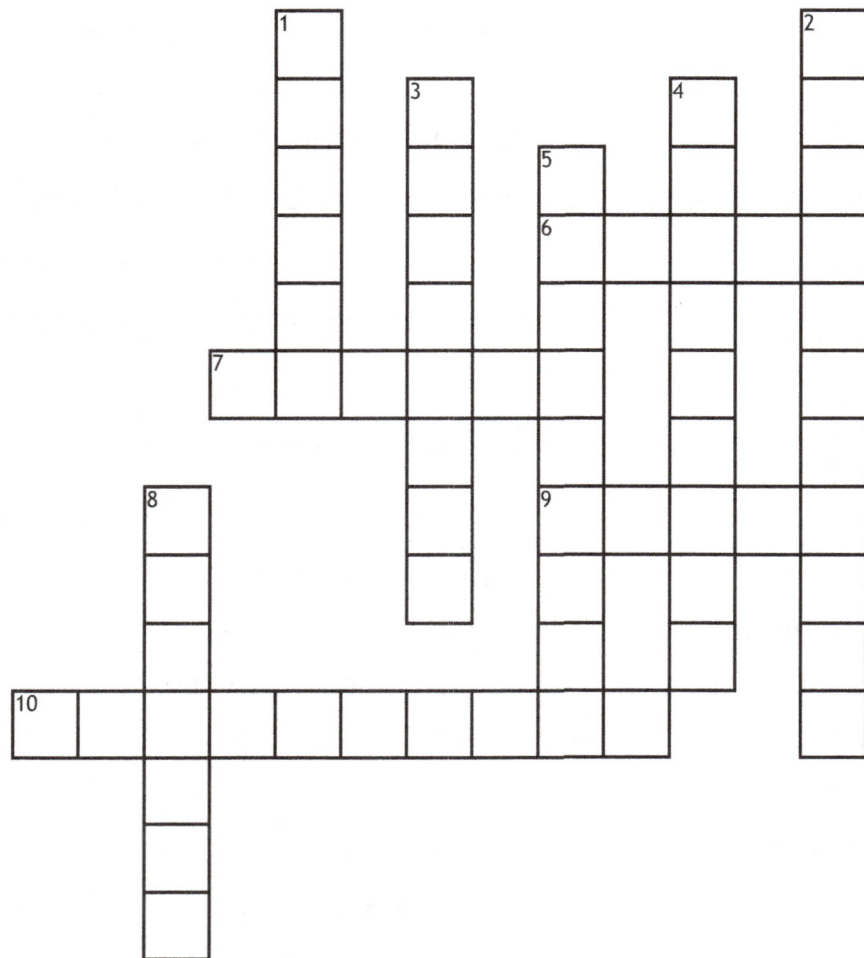

Down

1. The amount of space something takes up

2. A tool that measures temperature

3. How close a measurement is to the true value

4. To adjust a tool so it measures correctly

5. Getting the same measurement over and

8. To find the size, amount, or degree of something

Measurement and Tools

```
V  B  O  I  W  N  B  H  Z  K  P  Y  N  B  D  K  G
A  Z  L  M  E  T  R  I  C  P  A  H  L  B  G  H  Q
Q  U  I  F  N  Y  T  Q  H  J  S  X  S  T  M  D  R
R  E  V  S  C  J  L  T  P  O  X  B  V  H  F  L  C
Z  R  G  I  N  S  T  R  U  M  E  N  T  E  I  V  A
R  S  H  I  E  B  T  W  K  A  D  M  L  R  N  R  Z
Q  P  R  E  C  I  S  I  O  N  K  Y  Y  M  V  C  N
P  X  G  O  O  H  V  G  C  S  U  X  T  O  F  I  I
A  M  M  E  A  S  U  R  E  A  S  Z  H  M  Q  J  P
B  C  C  S  H  Y  P  F  N  U  L  C  X  E  J  J  T
S  I  C  K  G  N  M  M  D  P  R  I  S  T  D  Q  T
J  C  L  U  M  O  Q  T  L  V  U  E  B  E  O  Z  K
K  N  A  Z  R  A  L  T  Q  R  L  Y  Q  R  M  U  W
B  U  B  L  I  A  J  Z  D  Q  E  V  A  S  A  N  F
H  I  B  R  E  W  C  V  Z  U  R  I  L  L  J  T  F
D  J  A  R  J  C  X  Y  W  W  D  R  T  L  H  R  E
V  O  Y  F  V  O  L  U  M  E  W  U  S  B  T  J  O
```

INSTRUMENT   MEASURE   VOLUME

THERMOMETER   SCALE   RULER

METRIC   CALIBRATE   PRECISION

ACCURACY

Safety in engineering is all about making sure we follow rules and use protective gear to keep ourselves and others safe while we build and create!

Safety always comes first in engineering! Before starting any project, engineers look for possible **HAZARD** spots and make sure they have the right **PROTECTION**. They carefully follow safety **GUIDELINES** and **PROCEDURES** to prevent accidents.

Every engineer knows that proper safety **EQUIPMENT** is crucial. Wearing **GOGGLES** to protect your eyes is just as important as using the right tools for the job. They also know to **INSPECT** all tools and machines before using them to make sure everything is working safely.

Good engineers always remember to **VENTILATE** their workspace, especially when using materials that might have strong odors. They exercise **CAUTION** when trying new things and know exactly what to do in case of an **EMERGENCY**.
Whether you're building a robot, conducting an experiment, or creating something new, following safety rules isn't just smart - it's essential! Remember, the best engineers are the ones who work safely and help others do the same.

WORD SCRAMBLE

1	PNECTSI	
2	OGSGLEG	
3	IETOTPNROC	
4	ICOTUAN	
5	LESUNGDIEI	
6	PEORCRDUE	
7	TLNEIEATV	
8	MNPIQEUET	
9	DZRHAA	
10	GCNMYEEER	

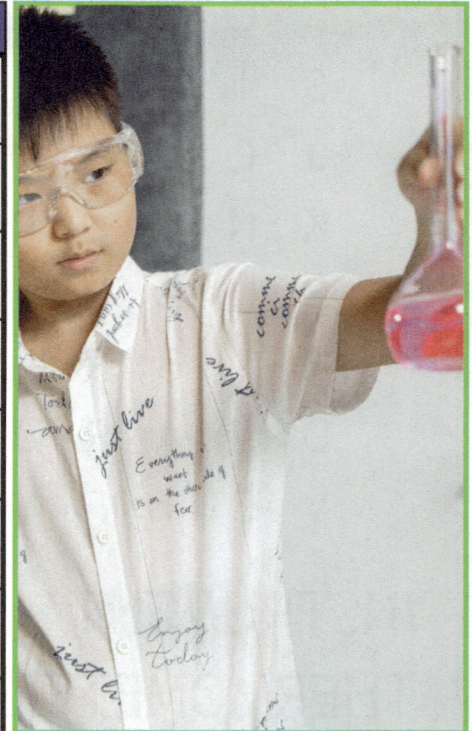

Safety in engineering means wearing goggles to protect our eyes and making sure we have good ventilation to keep the air fresh and safe while we work on our projects!

I. Key Definitions

Write the definitions for these important terms:

1. Hazard: _____

2. Protection: _____

3. Guidelines: _____

II. Fill in the Blank

Word Bank: CAUTION   EMERGENCY   EQUIPMENT   GOGGLES   INSPECT

1. Engineers need proper safety _____ for their work.

2. _____ protect your eyes during projects.

3. Engineers _____ tools before using them.

4. Engineers use _____ when trying new things.

5. Engineers know what to do in case of an _____.

III. True or False

Mark each statement True (T) or False (F):

1. _____ Safety comes last in engineering.

2. _____ Engineers should ventilate their workspace.

3. _____ Tools don't need to be checked before use.

4. _____ Following safety rules is essential.

IV. Matching

Match the safety actions with their purpose:

1. _____ Wear goggles   A. Clean air to breathe

2. _____ Inspect tools   B. Protect eyes

3. _____ Use caution   C. Work carefully

4. _____ Ventilate   D. Check for problems

V. Reflection

Why do you think it's important for engineers to help others work safely too? (Write 2-3 sentences)

_____

_____

_____

Safety in Engineering

Across

2. A set of steps to follow for safety

3. This is something that could cause harm or danger

9. To check something carefully for problems

10. Safety glasses that protect your eyes

Down

1. Rules that help keep people safe

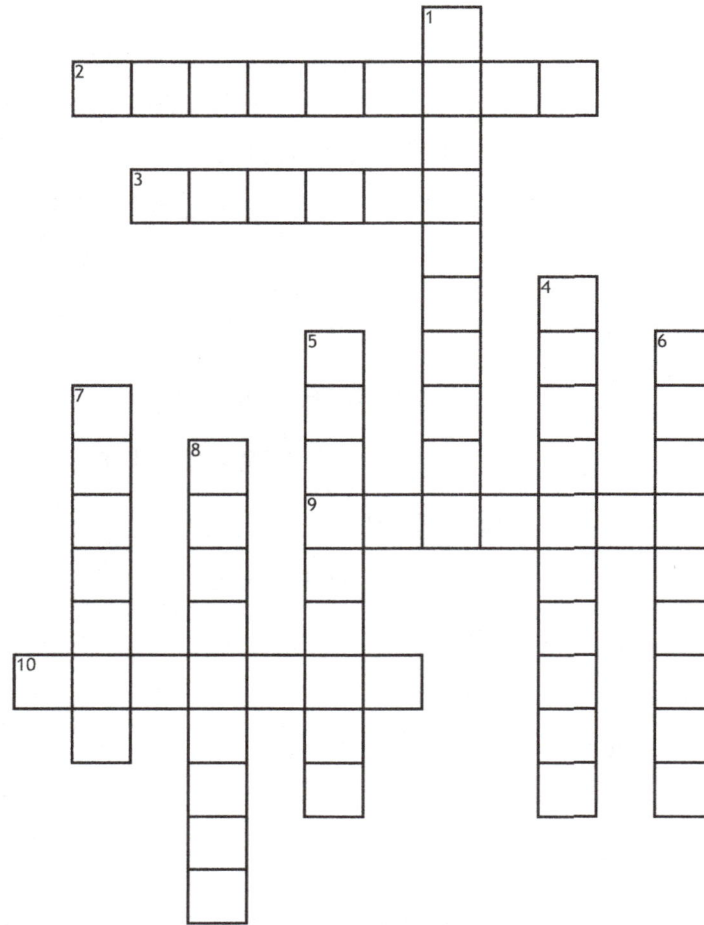

4. The equipment or actions that keep you safe

5. The tools and machines used for a task

6. To bring fresh air into a space

7. Being careful to avoid danger

8. A serious situation that needs quick action

Safety in Engineering

```
F  K  I  G  I  J  Q  M  N  G  M  P  M  U  H  L  Z
T  B  H  U  F  N  X  Y  B  J  Q  K  I  K  Z  Y  Q
H  L  O  I  B  A  S  X  T  N  R  Q  R  V  U  R  Q
L  Y  K  D  W  F  V  P  T  I  X  Q  F  Y  G  U  E
N  M  I  E  O  V  J  J  E  H  W  F  H  D  L  T  Q
N  P  G  L  J  W  E  Y  M  C  A  H  V  F  L  E  U
V  T  A  I  D  K  E  N  Y  V  T  Z  G  L  R  F  I
B  Z  F  N  C  Q  M  G  T  I  K  L  A  Z  C  C  P
H  O  G  E  I  Y  E  O  Z  I  Q  B  Z  R  O  V  M
C  T  Y  S  B  N  R  G  E  G  L  C  O  R  D  E  E
U  A  D  Y  M  J  G  G  X  X  E  A  H  Y  Z  Q  N
D  G  U  W  Q  F  E  L  G  M  R  P  T  I  D  I  T
H  Z  P  T  P  V  N  E  P  V  Y  D  X  E  G  S  R
D  N  W  I  I  U  C  S  P  X  F  X  S  N  M  Q  F
P  R  I  Y  O  O  Y  W  R  Q  Z  Z  E  K  R  S  M
Y  L  D  T  C  X  N  P  R  O  T  E  C  T  I  O  N
K  K  C  T  U  V  P  R  O  C  E  D  U  R  E  F  J
```

GUIDELINES   CAUTION   INSPECT
VENTILATE   EMERGENCY   GOGGLES
EQUIPMENT   PROCEDURE   PROTECTION
HAZARD

Structural engineering is all about designing buildings and bridges to make sure they can hold up against different forces, like weight and wind, so they stay strong and safe!

Have you ever wondered how tall buildings stay standing? It all starts with a strong **FOUNDATION** that spreads the building's **LOAD** across the ground. Engineers must think carefully about forces like **COMPRESSION** and **TENSION** when designing structures. **COMPRESSION** pushes things together, like when you squeeze a spring, while **TENSION** pulls things apart, like a stretched rubber band.

Every structure must fight against **GRAVITY**, which constantly pulls everything downward. That's why buildings need careful **SUPPORT** systems. **BEAMS** span spaces to hold up floors and ceilings, while **TRUSSES** use triangles to create super-strong frameworks. Some ancient buildings use clever designs like the **ARCH**, which spreads weight to the sides, and **BUTTRESSES**, which are like giant hands pressing against walls to keep them from falling over.

These simple but brilliant ideas have helped humans build amazing structures that have lasted for centuries!

WORD SCRAMBLE

1	SPMOOERSICN	
2	BESRSTTU	
3	RUSST	
4	MAEB	
5	ALDO	
6	INNTOSE	
7	TARVYIG	
8	CARH	
9	NAIOONUTDF	
10	TSOPURP	

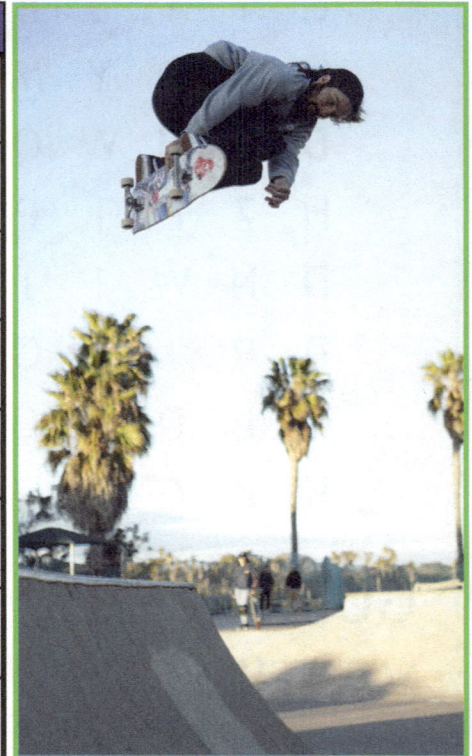

Gravity is the force that pulls everything toward the Earth, and understanding how it works helps us design strong structures that can stand up to its pull!

I. Key Definitions

Write the definitions for these important terms:

1. Foundation: _____

2. Compression: _____

3. Tension: _____

II. Fill in the Blank

Word Bank: ARCH   BEAMS   GRAVITY   LOAD   TRUSSES

1. A building's weight is called its _____.

2. _____ pulls everything downward.

3. _____ hold up floors and ceilings.

4. _____ use triangles to make strong frameworks.

5. An _____ spreads weight to the sides.

III. True or False

Mark each statement True (T) or False (F):

1. _____ Buildings don't need support systems.

2. _____ Compression pushes things together.

3. _____ Triangles help make structures stronger.

4. _____ Ancient buildings used clever designs.

IV. Matching

Match the structure parts with what they do:

1. _____ Foundation   A. Supports floors

2. _____ Beam   B. Holds up building

3. _____ Buttress   C. Uses triangles

4. _____ Truss   D. Supports walls

V. Reflection

Why do you think engineers need to think about different forces when designing buildings? (Write 2-3 sentences)

_____

_____

_____

Forces and Structures

Across

2. A curved structure that spreads weight to its sides

6. An external support that helps walls stand against pressure

8. The weight or force carried by a structure

9. A stretching force that pulls on both ends of an object

10. A framework of beams arranged in triangles for strength

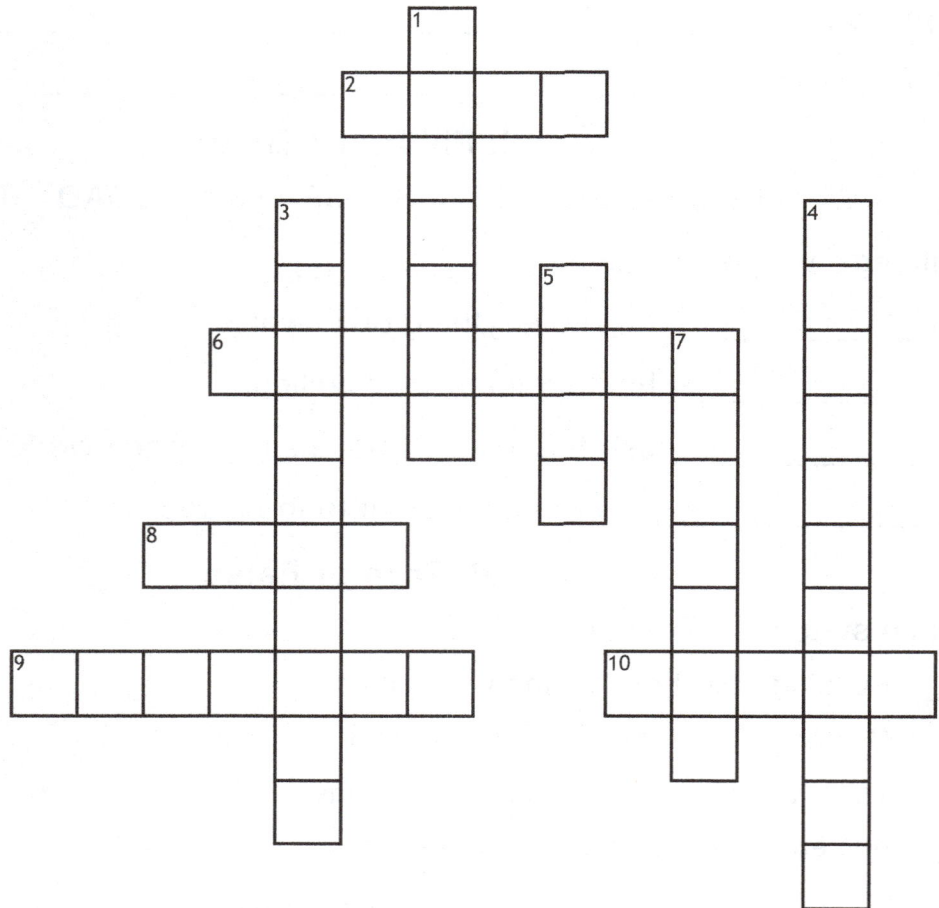

Down

1. The force that pulls everything toward Earth's center

3. The base that supports a building or structure

4. When a force pushes inward on both ends of an object

5. A long piece that spans a space to hold up weight above

7. When something that holds up or bears the weight of another object

Forces and Structures

```
R P Y E D N R P M A T K O L C T F
Q N Q I H X R P W A R T K N C Q D
J E M C H T J U M O I C W T O O T
F O U N D A T I O N G R H F M H E
S L G Z K W Z L Q I R T E T P U N
G T R U S S O Q M O V K L J R Y S
I X S U P P O R T P C I N Z E K I
P E C E R Z O Y C L O Z P B S H O
P R J N U A D Z B B M C E S S T N
G L N Q A U R W U W C O C N I F G
O R O M J N C Z T J H N B B O X R
P W A A D Q Z J T B G K Z W N K Y
B O N V D M F J R V E J V J Y Z K
K P V N I B M B E M T A Q I V L J
A V K P R T Q I S I U I M Z U G U
F S Q X T V Y P S T W P P H O S K
K Y D U F F C O Q P A P U A T L X
```

BUTTRESS   ARCH   TRUSS
BEAM   LOAD   SUPPORT
FOUNDATION   GRAVITY   TENSION
COMPRESSION

Building materials, like wood, metal, and concrete, are the different things we use to create strong and safe structures, each with its own special qualities!

The world around us is built from amazing materials! Modern builders often start with **CONCRETE**, a super-strong material that starts liquid but hardens like rock. They might add steel **REINFORCEMENT** to make it even stronger. Many new buildings use **COMPOSITE** materials that combine the best properties of different substances.

Metals called **ALLOYS**, like the steel in skyscrapers, are stronger than pure metals alone. **POLYMER** materials, like the plastic in your water bottle, are lightweight yet tough. **CERAMIC** tiles protect our homes and can handle extreme heat. Good **INSULATION** keeps buildings warm in winter and cool in summer.

Builders use special **ADHESIVES** to stick materials together permanently. **TIMBER** remains popular because it's both beautiful and strong. The **DURABILITY** of these materials is important - nobody wants their house falling apart! By choosing the right materials, engineers can create buildings that last for generations while keeping people safe and comfortable.

WORD SCRAMBLE

1	AEIESHVD	
2	RELYMPO	
3	LAOLY	
4	IOUNILATSN	
5	ONEMFRTNIECER	
6	RUDLAYTIIB	
7	CIOPOTSME	
8	EBMTRI	
9	EETOCRNC	
10	RAIECCM	

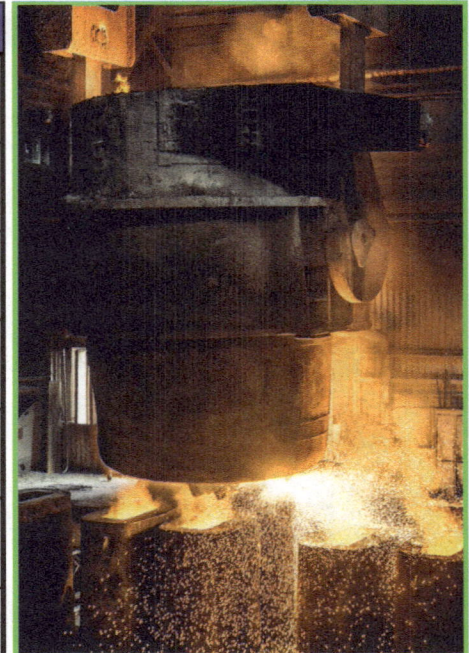

Building materials include alloys, which are strong metals made from mixing different metals, polymers, which are flexible and lightweight materials like plastic, and ceramics, which are hard and heat-resistant, each helping us create amazing structures!

I. Important Definitions

Write the definition for each term using the article:

1. Concrete: _____

2. Alloys: _____

3. Durability: _____

II. Fill in the Blanks

Word Bank: POLYMER   CERAMIC   INSULATION   TIMBER   ADHESIVES

1. _____ tiles can handle extreme heat and protect our homes.

2. _____ keeps buildings warm in winter and cool in summer.

3. _____ materials, like plastic water bottles, are lightweight but tough.

4. _____ is used to stick materials together permanently.

5. _____ is popular because it's both beautiful and strong.

III. True or False

Mark each statement as True or False:

1. _____ Concrete starts as a liquid but hardens like rock.

2. _____ Steel reinforcement makes concrete weaker.

3. _____ Composite materials combine properties of different substances.

4. _____ Plastic materials are very heavy.

IV. Matching

Match the material with its correct description. Write the letter on the line:

1. _____ Concrete   A. Beautiful and strong natural material

2. _____ Steel   B. Starts liquid, hardens like rock

3. _____ Timber   C. Used in skyscrapers

4. _____ Adhesives   D. Sticks materials together

V. Reflection

Why do you think it's important for builders to choose the right materials?

_____

_____

_____

Building Materials

Across

2. Material that prevents heat from moving in or out

5. A substance that bonds materials together

7. A mixture of cement, sand, and water that hardens like rock

9. A metal made by mixing two or more different metals together

10. Material added to make something stronger

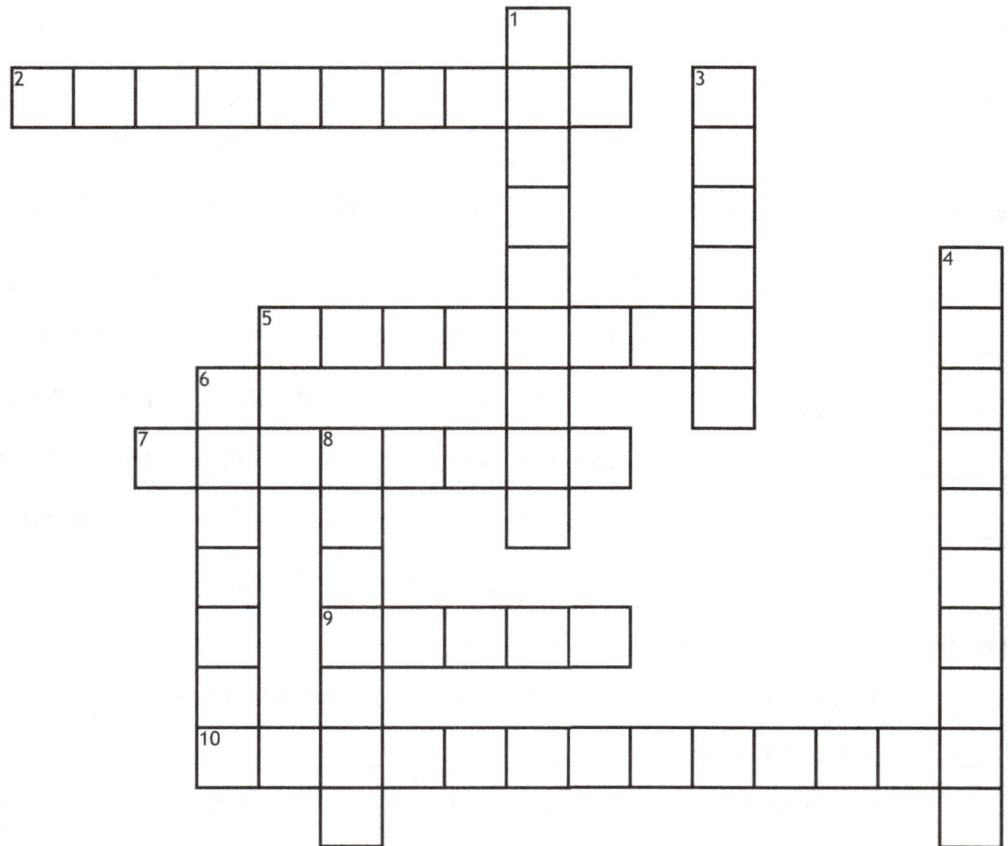

Down

1. A material made by combining two or more different materials

3. Wood prepared for use in building and carpentry

4. How long a material can last without breaking down

6. A material made of long chains of repeating molecules

8. A hard material made by heating clay at very high temperatures

Building Materials

```
W V X D C C S S U V S T L K V T J
V J Z R U O I H I B J N D K U V M
F E C E X R N N Z B O Q V V P L E
B A E I I B A C S Z Q E H P O B P
Q R R N M H F B R U M L G J L Z P
L D A F W S J I I E L P B W Y G C
H S M O Q Y X L H L T A N A M P U
E S I R T I M B E R I E T S E C Y
M O C C H L I H Q W V T K I R E C
O G C E K Q Q A R Z R Y Y R O K Y
N A T M W J G H D R X C I K E N B
D N M E D B T J Q H A G B I Q M D
C R M N E M K O I T E A M Z G H E
C R C T L M Q G J O L S L S Y A F
U R M V N Y J M K Y Q N I L M M V
O W U A C O M P O S I T E V O S T
C T Z S Z N Q C L O I H F E E Y L
```

DURABILITY   REINFORCEMENT   TIMBER

ADHESIVE   INSULATION   CERAMIC

POLYMER   ALLOY   COMPOSITE

CONCRETE

Bridge design is the process of planning and building bridges that are strong and safe, allowing people and vehicles to cross over rivers and valleys!

Bridges are amazing structures that help us cross rivers, valleys, and canyons! Each bridge's **SPAN** determines what type it needs to be. For very long distances, engineers often choose **SUSPENSION** bridges, which use strong **CABLES** hanging from tall towers to hold up the bridge **DECK** where cars drive.

Every bridge needs solid **ABUTMENTS** where it meets the land, and many have **PIERS** in the water to provide extra support. Large **GIRDERS** run along the length of the bridge to keep it strong. In suspension bridges, massive **ANCHOR** blocks hold the cables firmly in place.

Some bridges use **CANTILEVER** designs, where sections stick out like diving boards. Others use ancient arch designs, where a **KEYSTONE** at the top locks everything together. Whether you're walking across a small footbridge or driving across the Golden Gate Bridge, these clever engineering solutions work together to keep you safe above the water!

WORD SCRAMBLE

#	Word	
1	KECD	
2	SBEACL	
3	ANRCIELVET	
4	NCHRAO	
5	PRIE	
6	YETENKOS	
7	APNS	
8	NESNPUIOSS	
9	DIRREG	
10	EMATBNUT	

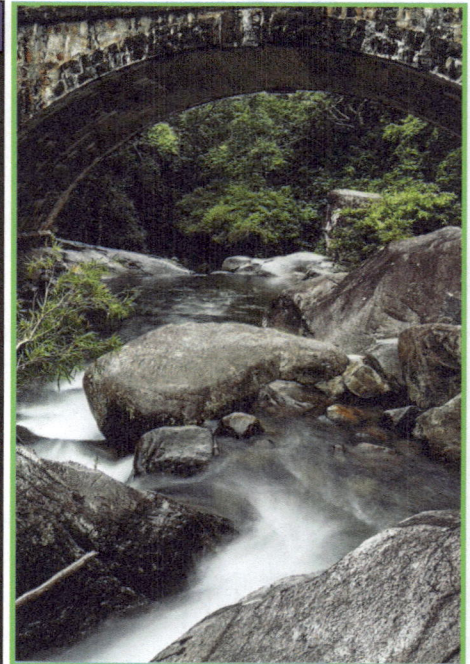

Cantilever design uses beams that extend out without support underneath, while keystone design focuses on a central stone at the top of an arch that holds everything together, making both structures strong and stable!

I. Important Definitions

Write the definition for each term using the article:

1. Span: _____

2. Suspension Bridge: _____

3. Deck: _____

II. Fill in the Blanks

Word Bank: CABLES   PIERS   GIRDERS   ABUTMENTS   KEYSTONE

1. _____ hang from tall towers to hold up suspension bridges.

2. _____ provide extra support in the water.

3. _____ are where the bridge meets the land.

4. _____ run along the length of the bridge to keep it strong.

5. In arch bridges, the _____ at the top locks everything together.

III. True or False

Mark each statement as True or False:

1. _____ Suspension bridges are used for very long distances.

2. _____ Cantilever bridges stick out like diving boards.

3. _____ Bridge cables are held in place by anchor blocks.

4. _____ The arch is a new type of bridge design.

IV. Matching

Match the bridge part with its correct description. Write the letter on the line:

1. _____ Deck   A. Provides support in water

2. _____ Cables   B. Where cars drive

3. _____ Piers   C. Hold up suspension bridges

4. _____ Abutments   D. Where bridge meets land

V. Reflection

Why do you think engineers need to use different types of bridges for different situations?

_____

_____

_____

Bridge Design

Across

2. The distance a bridge crosses from one support to another

5. The center stone that locks an arch bridge together

7. A vertical support that holds up a bridge between its ends

9. A beam supported at only one end

10. A large beam that supports the bridge deck

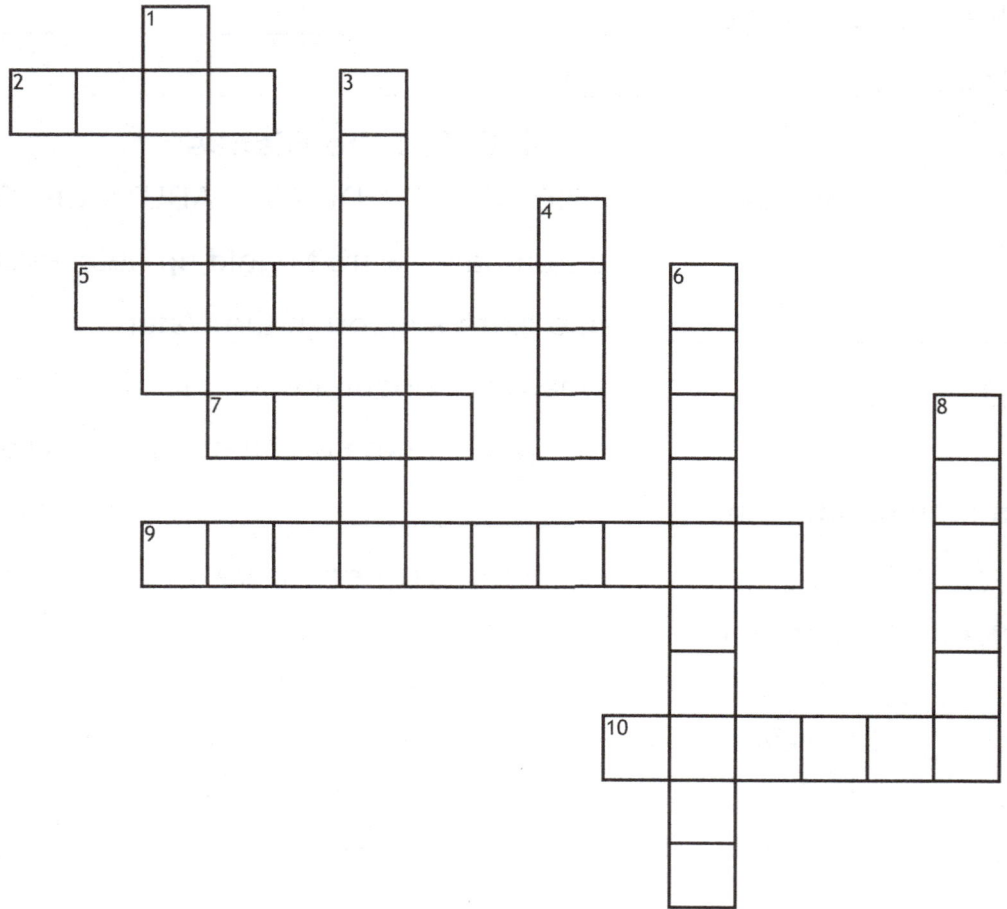

Down

1. Strong metal ropes that hold up suspension bridges

3. The end supports of a bridge that connect it to land

4. The flat surface of a bridge that traffic moves across

6. A type of bridge that hangs from strong cables

8. A structure that holds suspension bridge cables in place

Bridge Design

```
O K I E N C A B L E S N F D L G O
G D P S U S P E N S I O N W X Y P
O Z H Y B L A D U A M Y V P X E F
R E P T Q S Q M P C N X A U T N B
A J V C U F Q T F F I C F I W K W
J Q Y X A T A G M Z Q H H G D Z N
X P S I O N L V Y K H C B O V O K
M R P O A G T B N L T E X Z R Q Y
K E A Q Y B S I T M Z K E D D E N
G Z N D H J U K L E F R X B T Z Q
G I R D E R P T O E U J B N B I X
V S E L M N C I M K V K Z B M U O
R Y K E Y S T O N E F E H R D Z S
T C M C D G I R C P N Z R J E D Y
T C X F X E A H Z J I T W L C O V
I R M C O U F S T B X E L C K F A
Z B C U O N W T F X H M R X K P B
```

KEYSTONE   CANTILEVER   ANCHOR

DECK   GIRDER   CABLES

PIER   ABUTMENT   SUSPENSION

SPAN

Architecture basics involve designing buildings and spaces that are not only beautiful but also functional and safe for people to use!

Before construction begins, architects create detailed **BLUEPRINTS** to show builders exactly what to do. They think carefully about **SCALE** and **PROPORTION** to make sure everything fits together perfectly. The **LAYOUT** of the building determines how people will move through its spaces.

When you look at a building's **EXTERIOR**, you might notice its **FACADE** has perfect **SYMMETRY**, with both sides matching like a mirror image. Tall **COLUMNS** might line the entrance, making it look grand and strong. The **INTERIOR** spaces are designed to be both beautiful and practical, with each room flowing naturally into the next.

Architects draw **ELEVATION** views to show how the building will look from different sides. They must consider everything from the height of door handles to the slope of the roof. Whether designing a cozy home or a towering skyscraper, architects blend art and science to create buildings that are both beautiful and useful!

WORD SCRAMBLE

1	RXOIEERT	
2	TRNIBPUEL	
3	LECSA	
4	NETLVAOEI	
5	OMNUCL	
6	ONRPOORPTI	
7	CAEDAF	
8	SMTYERMY	
9	OAYUTL	
10	RERNIOIT	

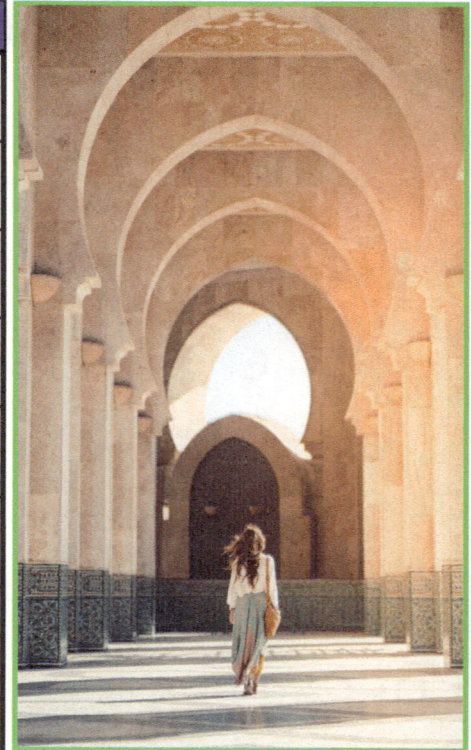

The exterior of a building is its outside appearance, or facade, which often features symmetrical designs and columns that add strength and beauty!

I. Important Definitions

Write the definition for each term using the article:

1. Blueprints: _____

2. Symmetry: _____

3. Elevation Views: _____

II. Fill in the Blanks

Word Bank: FACADE   LAYOUT   COLUMNS   EXTERIOR   INTERIOR

1. The _____ of a building determines how people move through its spaces.

2. When you look at a building's _____, you can see how it looks from the outside.

3. Tall _____ might line the entrance to make it look grand.

4. The building's _____ is the front face that people see first.

5. _____ spaces are designed to be both beautiful and practical.

III. True or False

Mark each statement as True or False:

1. _____ Architects create blueprints after construction begins.

2. _____ Symmetry means both sides match like a mirror image.

3. _____ Rooms should flow naturally from one to the next.

4. _____ Architects only need to think about big things like roofs.

IV. Matching

Match the architectural term with its correct description. Write the letter on the line:

1. _____ Blueprint   A. Outside of the building

2. _____ Exterior   B. Shows how to build it

3. _____ Layout   C. Inside spaces

4. _____ Interior   D. How people move through spaces

V. Reflection

Why do you think it's important for architects to plan everything carefully before construction begins?

_____

_____

_____

Architecture Basics

Across

3. The way spaces are arranged in a building

5. A drawing showing how a building looks from the side

6. The outside surfaces of a building

8. A detailed plan showing how to build a structure

9. When both sides of a building mirror each other

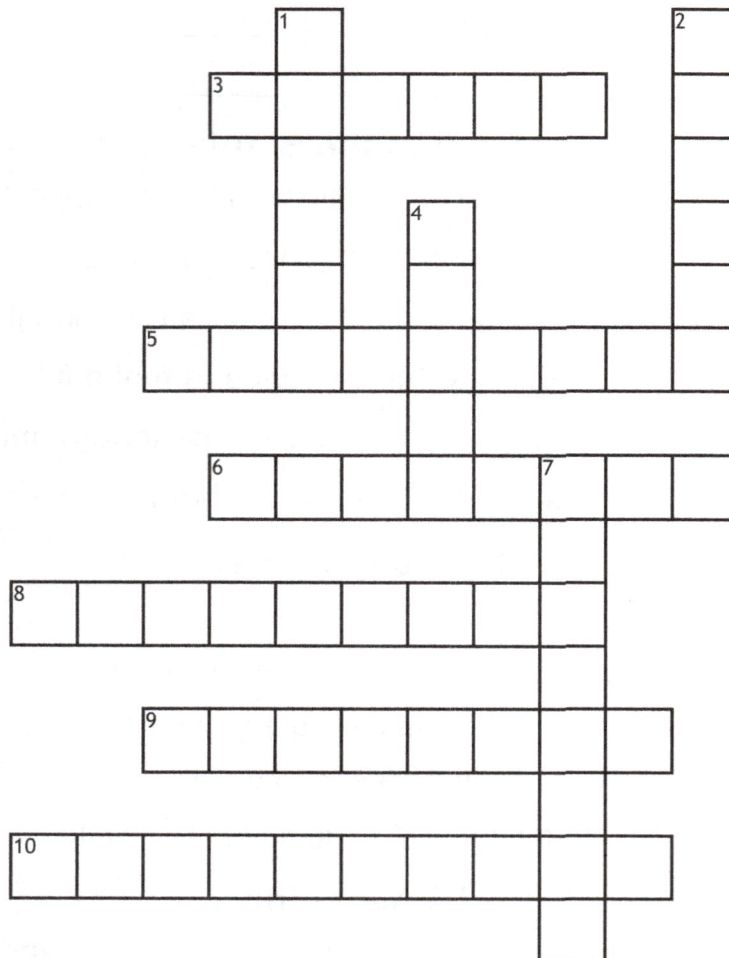

10. How different parts of a building relate in size

Down

1. The front face or exterior wall of a building

2. A vertical pillar that helps support a structure

4. The size relationship between parts of a building

7. The inside spaces of a building

Architecture Basics

```
S T B T B C O L U M N U T G O K G
P L L Y W V W A G M R F E D K B P
Y A S M J F M X Q E X T E R I O R
Q T V X P P Q I E Y F W V W F S O
K F A M R L A V B L I Q O Q F G Q
K Y M O O G H R G L E T N F F B B
W S Y Q P R B M F C M V M D H L G
Q S S Z O X H H Q J I O A N E Q C
F Y G E R H M K I G S H D T B Z Q
A M R V T Z F Y V N T S W A I Q J
C M L Y I A W Q H W T W S Q J O W
Q E C U O J F H Q M S E V C J M N
M T R Y N A A F B O Q V R M A R P
C R H D K G P L A Y O U T I M L N
A Y V X F A C A D E I N I I O D E
G B R J M D G F Z T P J M R L R E
J Z F Y Q T B L U E P R I N T I J
```

LAYOUT   PROPORTION   ELEVATION
EXTERIOR   INTERIOR   SCALE
COLUMN   FACADE   SYMMETRY
BLUEPRINT

Earthquake-resistant structures are specially designed buildings that can move and flex during an earthquake, helping to keep people safe and prevent the buildings from collapsing!

When **SEISMIC** activity strikes, buildings need special designs to stay standing. **SHOCKWAVES** travel through the ground, shaking buildings from their **FOUNDATION** up. Smart engineers make buildings FLEXIBLE so they can sway without breaking. They also install special **DAMPER** systems that work like shock absorbers in a car.

RESONANCE is a dangerous effect that can make buildings shake more violently during earthquakes. To prevent this, engineers carefully design structures to avoid matching the earthquake's motion. Many older buildings need a **RETROFIT** to add earthquake protection. Special **JOINT** connections let different parts of the building move separately, increasing **STABILITY** during shaking.

Strong **REINFORCEMENT** in walls and columns helps hold everything together when the ground moves. Modern earthquake-resistant buildings might look normal from the outside, but they're filled with clever engineering that helps them dance with the earthquake instead of fighting against it!

WORD SCRAMBLE		
1	SIATILTBY	
2	PMADRE	
3	SAKEOWCVH	
4	ENSECONRA	
5	INJOT	
6	TIENECENORFMR	
7	ELFILXEB	
8	RFOIETTR	
9	NIOUDNOTFA	
10	CSSIIEM	

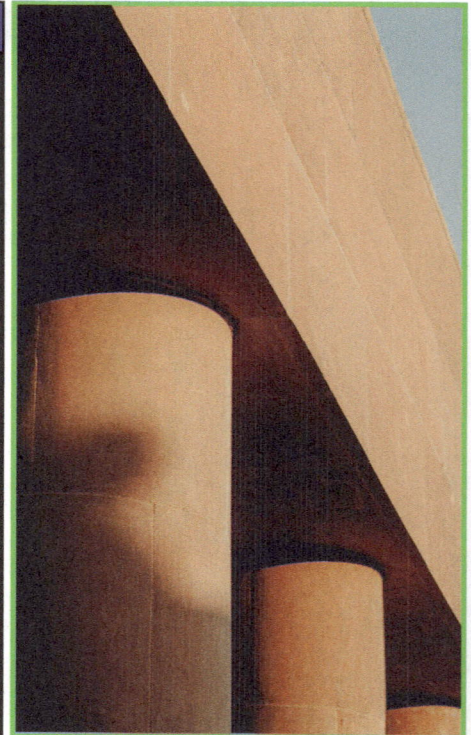

Concrete wall reinforcement for earthquakes involves adding extra materials, like steel bars or mesh, to make the walls stronger and better able to withstand shaking during an earthquake!

I. Important Definitions

Write the definition for each term using the article:

1. Seismic Activity: _____

2. Shockwaves: _____

3. Damper: _____

II. Fill in the Blanks

Use words from the word bank to complete each sentence:

Word Bank: FLEXIBLE   FOUNDATION   JOINT   STABILITY   REINFORCEMENT

1. Engineers make buildings _____ so they can sway without breaking.

2. Earthquakes shake buildings from their _____ up.

3. Special _____ connections let different parts of the building move separately.

4. _____ in walls and columns helps hold everything together.

5. Moving separately during shaking helps increase _____.

III. True or False

Mark each statement as True or False:

1. _____ Damper systems work like shock absorbers in a car.

2. _____ Buildings should try to match the earthquake's motion.

3. _____ Older buildings might need changes to add earthquake protection.

4. _____ Earthquake-resistant buildings always look very different from normal buildings.

IV. Matching

Match the earthquake term with its correct description. Write the letter on the line:

1. _____ Foundation   A. Lets building parts move separately

2. _____ Flexible   B. Bottom of the building

3. _____ Joint   C. Helps hold walls together

4. _____ Reinforcement   D. Can sway without breaking

V. Reflection

Why do you think it's important for buildings in earthquake zones to be able to move and sway?

_____

_____

_____

Earthquake-Resistant Structures

Across

5. Related to earthquakes and earth movements

7. An energy wave that moves through the ground

8. The materials added to make a structure stronger

10. To add new features to an older building

Down

1. A device that reduces shaking in buildings

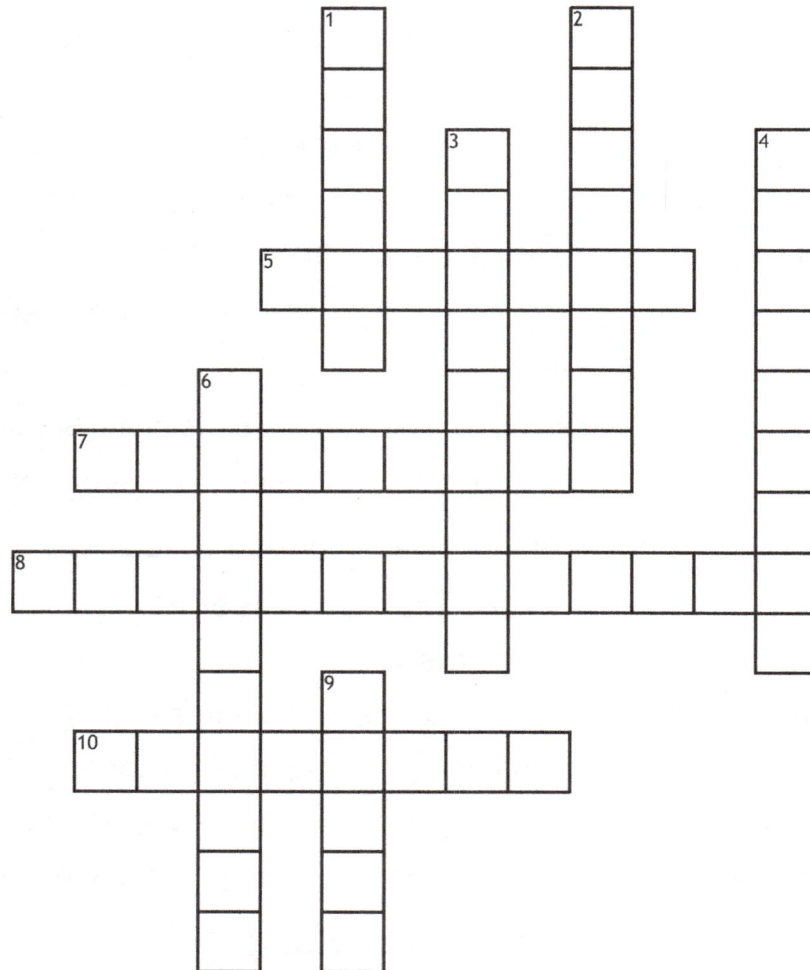

2. Able to bend without breaking

3. When shaking matches a building's natural motion

4. How well a structure resists tipping or swaying

6. The base system that anchors a building to the ground

9. A connection between building parts that can move

Earthquake-Resistant Structures

```
D N W Y Z O K S K M F X G B H S R
Q L R J O I N T T U I M J P P A E
J P O S T B C S V A K R F F P G T
B L R H E H D Z Y O B E X A X T R
M V F E Q I G Q T X L I M F J C O
C X L D S I S C Y Y F N L O P I F
V N E X V O Y M A T U F I I O E I
U D X F T B N F I S T O V G T V T
T V I O E P J A L C R R N J Z Y P
R H B U G N C Z N J W C R E H L H
B V L N T H G H J C N E V Q P V N
L F E D N H R P U U E M H T X H Y
W P G A D A M P E R O E B B H Z V
G Y B T F I S E G P K N F T I M D
Y T T I D R D N C J K T V H A W H
X H C O K T X S H O C K W A V E H
D V F N K C T X N Q B I H F B E X
```

REINFORCEMENT   STABILITY   FOUNDATION

JOINT   RETROFIT   RESONANCE

DAMPER   FLEXIBLE   SHOCKWAVE

SEISMIC

Simple machines are basic tools, like levers and pulleys, that help us make work easier by changing the direction or amount of force we use.

Simple machines are amazing tools that make our lives easier every day! When you use a see-saw at the playground, you're actually using a **LEVER** that moves around a **FULCRUM** to lift your friend up and down. **FORCE** is needed to make things move, but simple machines help us use less **WORK** to get jobs done.

Look at your bicycle - it uses a **WHEEL** and **AXLE** system to help you roll smoothly down the street. When you go up a hill, you're using an **INCLINE** to make climbing easier. Your bike chain runs through a **PULLEY** system to help turn the wheels with less effort.

Even something as simple as a door stopper is a **WEDGE** that holds the door open, while the threads on a **SCREW** help hold things together tightly. Simple machines are everywhere - once you start looking for them, you'll be amazed at how many you can find!

	WORD SCRAMBLE	
1	ULYPLE	
2	XLAE	
3	OKRW	
4	ILNENCI	
5	EVELR	
6	SECRW	
7	LFRCUMU	
8	GWDEE	
9	FCROE	
10	WLEHE	

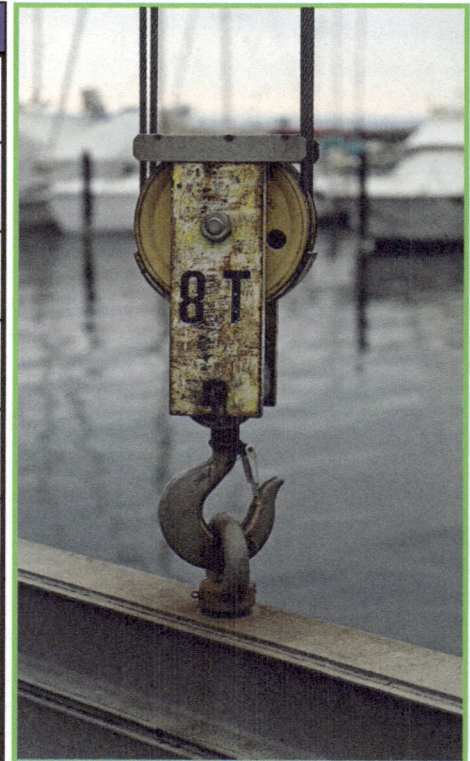

A pulley is a simple machine that uses a wheel and a rope to help lift heavy objects more easily by changing the direction of the force we apply.

I. Key Definitions

1. Force is: _____

2. Work is: _____

3. Fulcrum is: _____

II. Fill in the Blank

Word Bank: LEVER | WEDGE | PULLEY | WHEEL AND AXLE | INCLINE

1. A see-saw is an example of a _____.

2. A bicycle uses a _____ system to roll smoothly.

3. Going up a hill uses an _____ to make climbing easier.

4. A door stopper is a _____ that holds the door open.

5. A bike chain runs through a _____ system.

III. True or False

Mark each statement as True or False:

1. _____ Simple machines make our lives harder.

2. _____ A bicycle uses multiple simple machines.

3. _____ A screw helps hold things together tightly.

4. _____ Simple machines help us use less work to get jobs done.

IV. Matching

Match the simple machine to where you might find it:

1. ____ Lever   A. Door stopper

2. ____ Wheel and axle   B. Bicycle chain

3. ____ Wedge   C. See-saw

4. ____ Pulley   D. Bicycle wheel

V. Reflection

What is one way you use simple machines in your daily life? Explain how it helps you.

_____

_____

_____

Simple Machines

Across

4. A slanted surface that helps move objects up or down

5. A circular object that rolls to make moving easier

6. A triangular tool that can split things apart

8. A wheel with a rope that makes lifting easier

10. A push or pull that can make things move

Down

1. A spiral inclined plane wrapped around a cylinder

2. A rigid bar that moves around a fixed point to lift heavy objects

3. The rod or shaft that a wheel spins around

7. The pivot point where a lever turns

9. The energy used to move something from one place to another

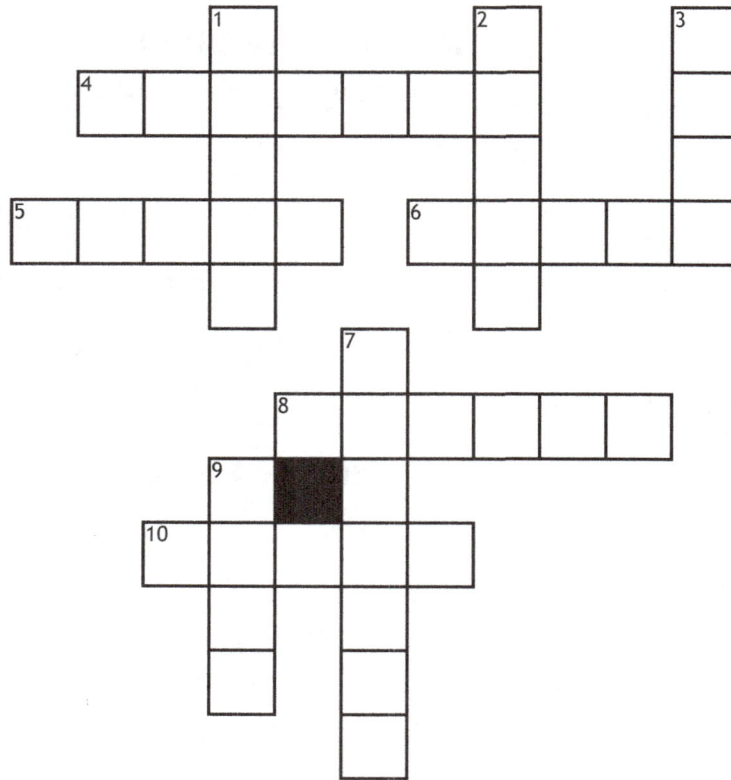

Simple Machines

```
P K Z Z B E W Q I N C L I N E Q N
F U W N O A Z S Q F W W D F T N C
D C K F O T W W K L V X P P H M F
B G O Z P U Q E V Q Y R Z Y N K Y
S C D L Q S U D O S W D U J I W G
G E J H P S P G K P C H U U G O Y
M C J E P G Q E O F U M E H W R P
J S H Z F X A L O I U L O E T K J
P Q A E N O K V I A O L L S L D U
D M L T T O R T V M M S C E P W A
H M E Z W J L C G I M Q Z R Y B U
S G V G S X M K E L P U M X U J O
E B E W T U X Z I A X L E R J M S
W B R R Y D P K L C O W J Q B I X
Q Q S L Y B B U M O E U P Z E I M
W S I Y Q M E X U O H O B J S Q Z
G M V O A S V A S C R E W Y B H U
```

WORK	FULCRUM	FORCE	INCLINE
AXLE	WHEEL	SCREW	WEDGE
PULLEY	LEVER		

Gears and pulleys are simple machines that work together to help us move things; gears turn to change the direction and speed of motion, while pulleys help lift heavy objects with less effort.

Have you ever wondered how your bicycle can go so fast? The secret lies in **GEARS** and pulleys! When you pedal, the **TEETH** of different-sized **GEARS** work together like tiny machines. The bigger the **DIAMETER** of a gear, the more power it can transfer.

A **PULLEY** system uses a **BELT** or rope to help lift heavy things or change the direction of **ROTATION**. You can find pulleys in many places, like flagpoles or elevator systems. When pulleys and gears work together in a **TRANSMISSION**, they can control **SPEED** and make work easier.

MECHANICAL systems need to be carefully designed to reduce **FRICTION**, which can slow things down and wear out moving parts. That's why bikes and machines need oil to keep their gears running smoothly. From simple toys to complex machines, gears and pulleys are amazing tools that help us do work more efficiently!

WORD SCRAMBLE

1	DEMEATRI	
2	AGER	
3	EHAMAICNCL	
4	SSIMROINSANT	
5	BTLE	
6	FRICINTO	
7	ELPUYL	
8	TAINOROT	
9	EEPDS	
10	ETEHT	

Mechanical systems are combinations of simple machines, like levers, gears, and pulleys, that work together to perform tasks and make our lives easier, like lifting, moving, or turning things.

I. Key Definitions

1. Diameter is: _____

2. Rotation is: _____

3. Friction is: _____

II. Fill in the Blank

Word Bank: TRANSMISSION | TEETH | BELT | GEARS | SPEED

1. When you pedal your bike, the _____ of gears work together.

2. Pulleys use a _____ or rope to lift heavy things.

3. _____ help your bicycle go fast.

4. A _____ can control speed and make work easier.

5. Pulleys and gears can control _____.

III. True or False

Mark each statement as True or False:

1. _____ Machines need oil to help reduce friction.

2. _____ The smaller the diameter of a gear, the more power it transfers.

3. _____ Pulleys can change the direction of rotation.

4. _____ Friction helps machines work faster.

IV. Matching

Match the item with where you might find it:

1. _____ Pulleys   A. Bicycle chain

2. _____ Gears   B. Elevator system

3. _____ Mechanical system   C. Flagpole

4. _____ Belt   D. Bicycle pedals

V. Reflection

Explain how gears or pulleys make one task in your life easier:

_____

_____

_____

Gears and Pulleys

Across

5. The distance across the center of a circle

7. Related to machines and their parts

8. A system that transfers power from one place to another

10. The pointed parts of a gear that lock together

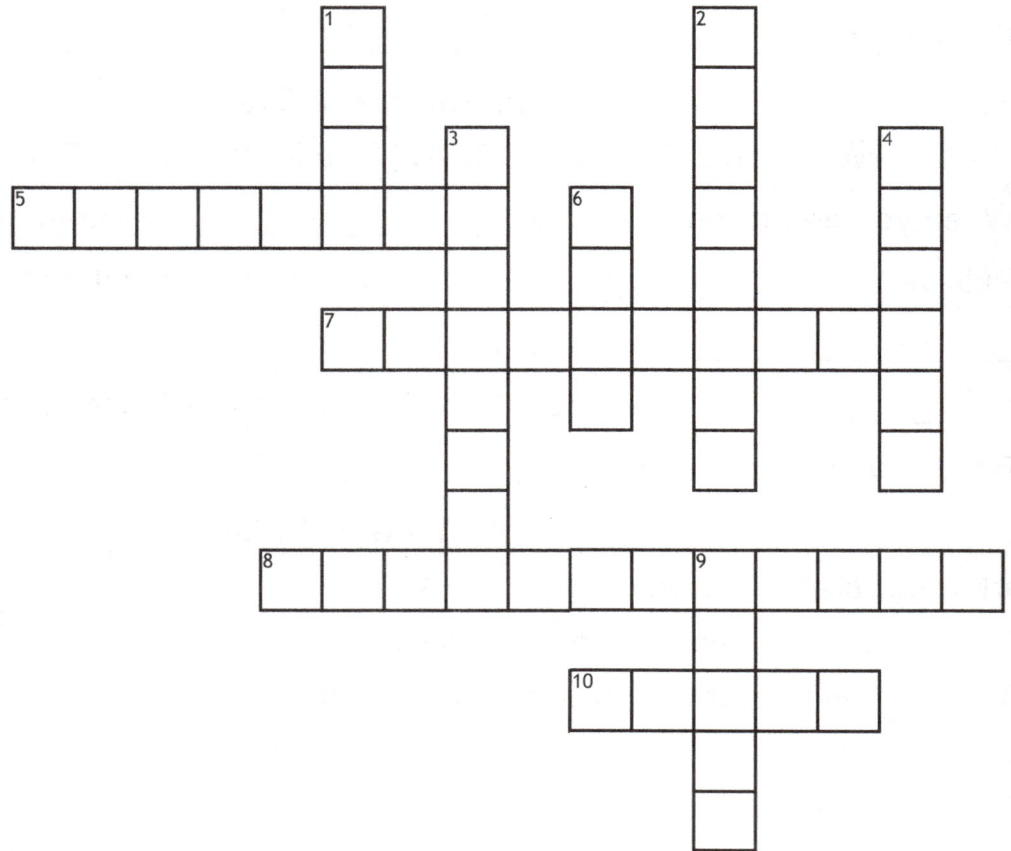

Down

1. A loop that connects pulleys to transfer power

2. The circular movement of an object around its center

3. A force that resists motion between surfaces

4. A grooved wheel that guides a rope or belt

6. A toothed wheel that meshes with others to transfer motion

9. How fast something moves or turns

Gears and Pulleys

```
U Z L X H J L Y P J I E Q E D Q H
R W W K V F Q I D P D V Q L U M L
L L Q M G O A A R R U F S Y Z N S
C R M T E E T H G M B L Q P G O T
U G G E R A N T B X M E L G E P R
A K T G J P R V D M E C L E Y E M
C R Y J L I K O U F C P P T Y C D
E U O R P T N E J R H D Q C T G V
F L A T N Z F U I I A A T O X T C
T W D A A G O B O C N Y G P Z C D
K X I Z Y T H V Z T I R D G J V S
N I A L M D I U O I C G W K K T E
F Q M I U U H O D O A X L R O F H
I B E I G E A R N N L P D E R U K
X Q T T R A N S M I S S I O N I U
E K E X W D X D A C B M P Y A W O
Q N R V V V Z Y H V E K J N B P J
```

TRANSMISSION   MECHANICAL   DIAMETER

FRICTION   SPEED   ROTATION

BELT   PULLEY   TEETH

GEAR

Motion and mechanics are about how things move and the forces that make them move, helping us understand why a ball rolls, a car drives, or a swing goes back and forth.

When you look at machines in **MOTION**, you're seeing **MECHANISMS** at work! A machine is like a team where each part has an important job. Some parts **PIVOT** around a fixed point, while others use **LINKAGE** systems to transfer movement from one place to another.

Have you ever seen how a **CAM** works? It's a fascinating part that can change circular motion into up and down movement. When parts **OSCILLATE** or **RECIPROCATE**, they create patterns of movement that can do amazing things, like running a sewing machine or moving windshield wipers.

KINETIC energy is found in all moving objects, and **MOMENTUM** helps keep them moving. When you ride a bike, **TORQUE** from your pedaling helps turn the wheels. Understanding how these mechanisms work helps engineers design better machines - from simple toys to complex robots. Motion and mechanisms are everywhere, making our world move in incredible ways!

WORD SCRAMBLE

1	OMINTO	
2	ITIKNEC	
3	ILNGAKE	
4	HECIAMMNS	
5	TOVPI	
6	UTQERO	
7	MAC	
8	MEMNUTMO	
9	RIEPETCCOAR	
10	TALOLESIC	

Cams and crankshafts work together in machines; cams turn round to push things up and down, while crankshafts help change that motion into circular movement, making everything run smoothly!

I. Key Definitions

Write the definition for each term:

1. Mechanism: _____

2. Kinetic Energy: _____

3. Torque: _____

II. Fill in the Blank

Use the words from the word bank to complete each sentence.

Word Bank: PIVOT | LINKAGE | OSCILLATE | CAM | MOMENTUM

1. When parts _____ around a fixed point, they help machines move.

2. A _____ can change circular motion into up and down movement.

3. _____ systems help transfer movement from one place to another.

4. When you ride a bike, _____ from your pedaling helps turn the wheels.

5. _____ helps keep moving objects in motion.

III. True or False

Mark each statement as True or False:

1. _____ A machine is like a team where each part has an important job.

2. _____ Engineers use their knowledge of mechanisms to design machines.

3. _____ Motion and mechanisms are only found in complex machines.

4. _____ Windshield wipers use reciprocating motion.

IV. Matching

Match the term with its description:

1. ___ CAM   A. Helps keep objects moving

2. ___ MOMENTUM   B. Changes circular motion to up and down

3. ___ RECIPROCATE   C. Moving back and forth

4. ___ MECHANISMS   D. Parts that make machines work

V. Reflection

Why do you think it's important for engineers to understand how different mechanisms work?

_____

_____

_____

Motion and Mechanisms

Across

6. Having to do with motion or movement

7. A turning or twisting force

8. A system of parts that work together in a machine

9. A rotating part that changes circular motion to up and down motion

10. A fixed point that something turns around

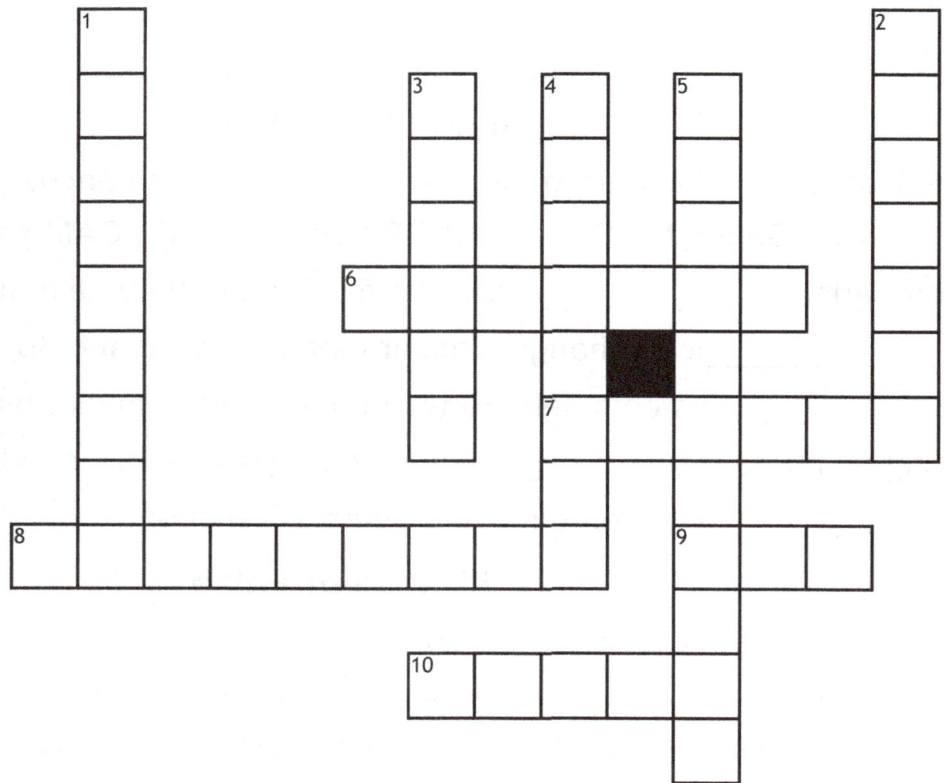

Down

1. To move back and forth regularly

2. A system of connected rods or bars that transfer movement

3. The act of moving from one place to another

4. The force of movement in a moving object

5. To move back and forth in a straight line

Motion and Mechanisms

```
M O M E N T U M O P L V D J X K L
M L P U L T W E C S D Z W O D Y I
N J R X L M O J U A C W Y H X U N
V B E U W F X R B I M I J K Q M K
W L C A S U X F Q B D J L O T B A
S V I Y K M P E K U U J O L R Y G
F B P M Y J Z I Z X E Q A B A M E
T D R P C K R H V S K X Z S R T F
Q I O S L J M P D O P I R Z K M E
K V C R C D E Q U P T J N L M K I
X V A O H O C Q N J D I W E P N Q
Q B T A A T H R U M Z O P H T C V
Z S E P W F A L I B O J S M E I C
M F R J A J N W F U V T A U W A C
F U U V V X I I R N X W I A R S O
W E N Y W R S X X H D D D O Q M G
J J G X B E M M M C Y F Y W N T Z
```

TORQUE   RECIPROCATE   KINETIC

MOMENTUM   OSCILLATE   CAM

LINKAGE   PIVOT   MECHANISM

MOTION

ENERGY is everywhere around us, working in amazing **SYSTEMS** that power our world! Think about a roller coaster - at the top of the hill, it has **POTENTIAL** energy just waiting to be used. As the cars rush down, that stored energy **CONVERTS** to **KINETIC** energy of motion. Energy can **TRANSFER** from one form to another, like when **SOLAR** panels change sunlight into electricity.

Energy systems are ways that we use different types of energy, like electricity or wind, to power machines and do work, helping us light our homes, move cars, and even play video games!

A **GENERATOR** is a clever machine that can turn motion into electrical energy, which then flows through a **CIRCUIT** to power our homes and devices. Even the heat you feel on a sunny day is **THERMAL** energy from the sun. Your body is an energy system too - it converts the food you eat into energy for running and playing!

Energy is never destroyed - it just changes form. From the smallest battery to the biggest power plant, energy systems help us do incredible things every day!

WORD SCRAMBLE

1	ELARHTM	
2	RIIUCTC	
3	RNERFSAT	
4	SALOR	
5	YSETSM	
6	LENAPOTIT	
7	NEVTROC	
8	RENGYE	
9	ROAGETREN	
10	KCIENIT	

Generators are machines that convert different types of energy, like wind or gasoline, into electrical energy, allowing us to power our homes, schools, and devices!

I. Key Definitions

Write the definition for each term:

1. Potential Energy: _____

2. Kinetic Energy: _____

3. Thermal Energy: _____

II. Fill in the Blank

Use the words from the word bank to complete each sentence.

Word Bank: SOLAR | TRANSFER | SYSTEMS | CIRCUIT | GENERATOR

1. Energy works in amazing _____ that power our world.

2. Energy can _____ from one form to another.

3. A _____ is a machine that turns motion into electrical energy.

4. _____ panels change sunlight into electricity.

5. Electrical energy flows through a _____ to power our homes.

III. True or False

Mark each statement as True or False:

1. _____ Energy can be destroyed completely.

2. _____ A roller coaster uses both potential and kinetic energy.

3. _____ Your body converts food into energy.

4. _____ Energy can change from one form to another.

IV. Matching

Match the term with its description:

1. ___ Potential Energy   A. Energy from the sun that feels hot

2. ___ Kinetic Energy   B. Energy waiting to be used

3. ___ Thermal Energy   C. Energy of motion

4. ___ Solar Panels   D. Changes sunlight to electricity

V. Reflection

Why do you think it's important to understand how energy works in our daily lives?

_____

_____

_____

Energy Systems

Across

2. Energy of motion in moving objects

8. Energy related to heat and temperature

9. The ability to do work or cause change

10. A path that electricity flows through

Down

1. A machine that produces electrical energy

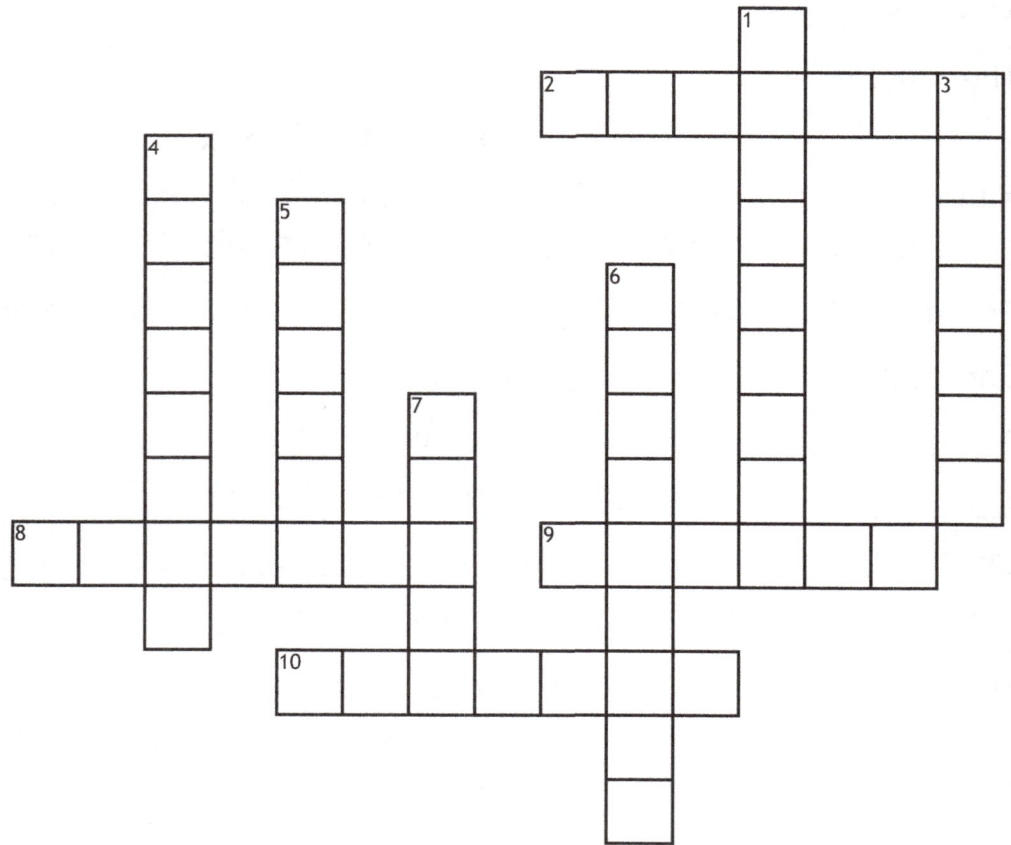

3. To change from one form of energy to another

4. The movement of energy from one place to another

5. A group of parts working together for a purpose

6. Stored energy that's ready to be used

7. Energy that comes from the sun

Energy Systems

```
J H K A M S K O P T D M I W R Q L
P F U R E A S G V Q D V G S F Y R
O O N R T G Y P E G A M Z C V N H
T G E L R V S S K T T H E R M A L
E S N P A E T N U S G F S Q H R S
N U E R N V E S J K O F R N R A H
T Z R Y S C M A C F Z L A M R V N
I W G I F Z F Z N I C G A Z N G O
A A Y A E I P S X B R T W R X W M
L P L F R T K W N P B C Z U P C C
W K K X G U J I O Q F C U B Y O V
H R T P A K V L N C M G S I J J Y
B C R R E M O R U E Y K L A T P F
W G E N E R A T O R T F M E W W P
J X K K C O N V E R T I W Z B Q A
C L M H B F M G V I G V C C T E R
F R G O A V Y K K Z S N W Q G N Y
```

GENERATOR   CIRCUIT   SOLAR

SYSTEM   CONVERT   THERMAL

KINETIC   POTENTIAL   TRANSFER

ENERGY

Transportation engineering involves creating vehicles that use engines, turbines, or other systems to generate thrust, which helps them move quickly and smoothly, like how airplanes soar through the sky or cars zoom down the road!

Transportation engineering helps us build amazing **VEHICLES** that can travel on land, water, and in the air! Every **VEHICLE** needs some form of **PROPULSION** to move forward. Airplanes use **TURBINES** and **THRUST** to fly through the sky, while their **AERODYNAMIC** shape helps them cut through the air with less resistance.

Cars and trucks have a strong **CHASSIS** that holds everything together, and a **SUSPENSION** system that makes the ride smooth over bumpy roads. Many modern vehicles use **HYDRAULIC** systems in their brakes and steering to make them safer and easier to control.

A powerful **LOCOMOTIVE** can pull heavy train cars along railroad tracks, using special wheels and engines. Modern transportation also relies on **NAVIGATION** systems to help drivers and pilots find the best routes to their destinations. From bicycles to spaceships, transportation engineers are always working to make vehicles better, faster, and safer!

WORD SCRAMBLE

1	EONUSSISNP	
2	ILMOOOVTEC	
3	RYDAUIHCL	
4	TTSHRU	
5	PNSLUPOORI	
6	HSSASIC	
7	MOACEYDIARN	
8	CIHELVE	
9	URTEBIN	
10	ITANVIGNAO	

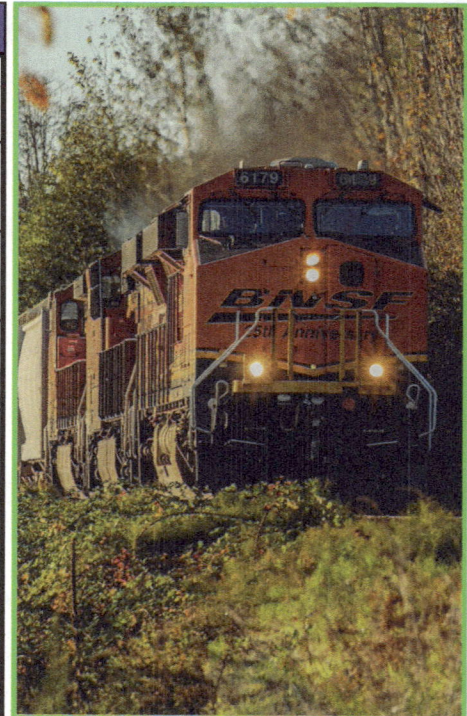

Transportation engineering helps design locomotives, which are powerful trains that pull cars full of people or goods, making it easier for us to travel long distances on tracks!

I. Key Definitions

Write the definition for each term:

1. Propulsion: _____

2. Chassis: _____

3. Aerodynamic: _____

II. Fill in the Blank

Use the words from the word bank to complete each sentence.

Word Bank: SUSPENSION | TURBINES | HYDRAULIC | NAVIGATION | LOCOMOTIVE

1. A _____ can pull heavy train cars along railroad tracks.

2. The _____ system makes the ride smooth over bumpy roads.

3. Airplanes use _____ to fly through the sky.

4. Modern vehicles use _____ systems in their brakes and steering.

5. _____ systems help drivers and pilots find the best routes.

III. True or False

Mark each statement as True or False:

1. _____ Vehicles can only travel on land.

2. _____ Every vehicle needs some form of propulsion to move.

3. _____ Transportation engineers work to make vehicles safer.

4. _____ Hydraulic systems make vehicles harder to control.

IV. Matching

Match the term with its description:

1. ___ Chassis   A. Helps cut through air

2. ___ Aerodynamic   B. Frame that holds everything together

3. ___ Navigation   C. Helps find the best route

4. ___ Turbines   D. Helps planes fly through sky

V. Reflection

Why do you think transportation engineering is important for our daily lives?

_____

_____

_____

Transportation Engineering

Across

2. The force that pushes something forward

5. A machine that moves people or things from place to place

7. A strong pushing force that moves something forward

8. The process of planning and following a route

10. An engine that pulls train cars

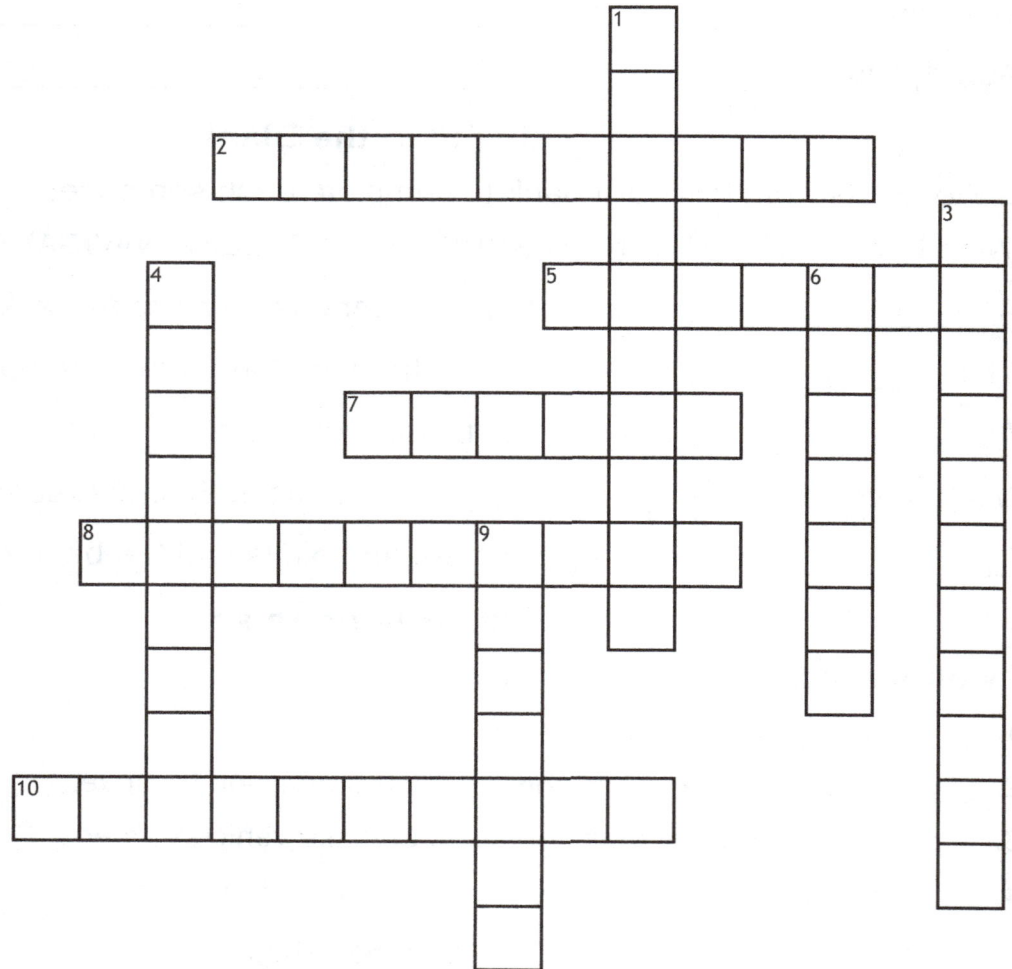

Down

1. System that absorbs shock and bumps while moving

3. Shaped to move smoothly through air

4. Using liquid pressure to move or lift things

6. The frame that supports a vehicle's parts

9. An engine that uses spinning blades to create power

Transportation Engineering

```
J N C D Y U O R Q M V K J B Q O K
S F P E G N N J D C X P Q S N P E
A V X R G Q A S U S P E N S I O N
S E L U O N H V P O N V P I S J R
X H C J M P C Y I E O A I V S M M
T I J P I V U U D G G J G L V B A
J C L L C U V L K R A K F F C D E
P L V A N C A L S Y A T X F O M R
T E N X P A A X W I C U I E W V O
V X J F K K K F R Y O Y L O E W D
H W U P F O U K I T Q N J I N R Y
H Z I L O C O M O T I V E N C R N
C H A S S I S F C A P G S I H C A
U L H Q J Y E Q Q B I Z U M P O M
H O P Y O H U B N S W U F I M O I
X F O T U R B I N E G H V Q B X C
Z J Q Y C Y R K T H R U S T K P V
```

NAVIGATION   LOCOMOTIVE   HYDRAULIC

SUSPENSION   THRUST   TURBINE

CHASSIS   AERODYNAMIC   PROPULSION

VEHICLE

Water systems are important networks that help collect, treat, and deliver clean water to our homes and schools, while also managing stormwater and wastewater to keep our environment healthy!

Water is always moving through Earth's amazing water systems! When the sun warms water, **EVAPORATION** occurs, turning liquid water into invisible water vapor. As this vapor rises and cools, **CONDENSATION** happens, forming clouds. These clouds release **PRECIPITATION** as rain or snow.

Some of this water becomes **RUNOFF**, flowing over land into streams. Small streams called **TRIBUTARIES** join together to form bigger rivers. This entire area of land is called a **WATERSHED**. Some water collects in **WETLANDS**, which are special areas where water covers the ground and creates homes for many plants and animals.

Other water soaks deep underground, becoming **GROUNDWATER** stored in an **AQUIFER** made of rock and soil. People often pump this water up for drinking and farming. We also build **RESERVOIRS** to store water for cities and towns.

Understanding these water systems helps us protect Earth's most precious resource - water!

WORD SCRAMBLE

1	OEESRIVRR	
2	TDAEWLN	
3	TNGRAUDREOW	
4	IAICITOTPEPNR	
5	TONOCINDSNEA	
6	WTEDHEARS	
7	VRTIOPNOEAA	
8	NFOURF	
9	RBTIARUTY	
10	UFEAIQR	

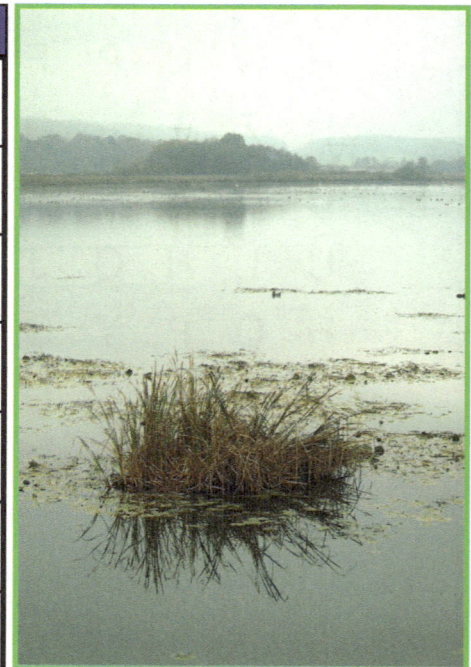

Water systems work with wetlands, which are special areas filled with water that help filter pollutants, provide homes for plants and animals, and store water to prevent flooding, making them crucial for a healthy environment!

I. Key Definitions

Write the definition for each term:

1. Evaporation: _____

2. Precipitation: _____

3. Watershed: _____

II. Fill in the Blank

Use the words from the word bank to complete each sentence.

Word Bank: RUNOFF | WETLANDS | GROUNDWATER | TRIBUTARIES | AQUIFER

1. Small streams called _____ join together to form bigger rivers.

2. Water that flows over land into streams is called _____.

3. _____ are special areas where water covers the ground and creates homes for plants and animals.

4. _____ is water stored deep underground.

5. An _____ is made of rock and soil and stores water underground.

III. True or False

Mark each statement as True or False:

1. _____ Condensation forms clouds in the sky.

2. _____ Water vapor is visible in the air.

3. _____ People pump groundwater for drinking and farming.

4. _____ Reservoirs store water for cities and towns.

IV. Matching

Match the term with its description:

1. ___ Evaporation   A. Water falling as rain or snow

2. ___ Condensation   B. Water turning into vapor

3. ___ Precipitation   C. Water vapor cooling to form clouds

4. ___ Wetlands   D. Areas where water covers the ground

V. Reflection

Why is it important to understand and protect Earth's water systems?

_____

_____

_____

Water Systems

Across

6. Water stored beneath Earth's surface

8. Large natural or human-made lake storing water

9. Smaller stream that flows into a larger river

10. Water that flows over land into streams and rivers

Down

1. When water vapor turns into liquid water droplets

2. Land area that drains water into a river system

3. Water falling from clouds as rain, snow, or hail

4. An underground layer that holds water in rock and soil

5. Process of water changing from liquid to gas

7. Area where land is covered by shallow water

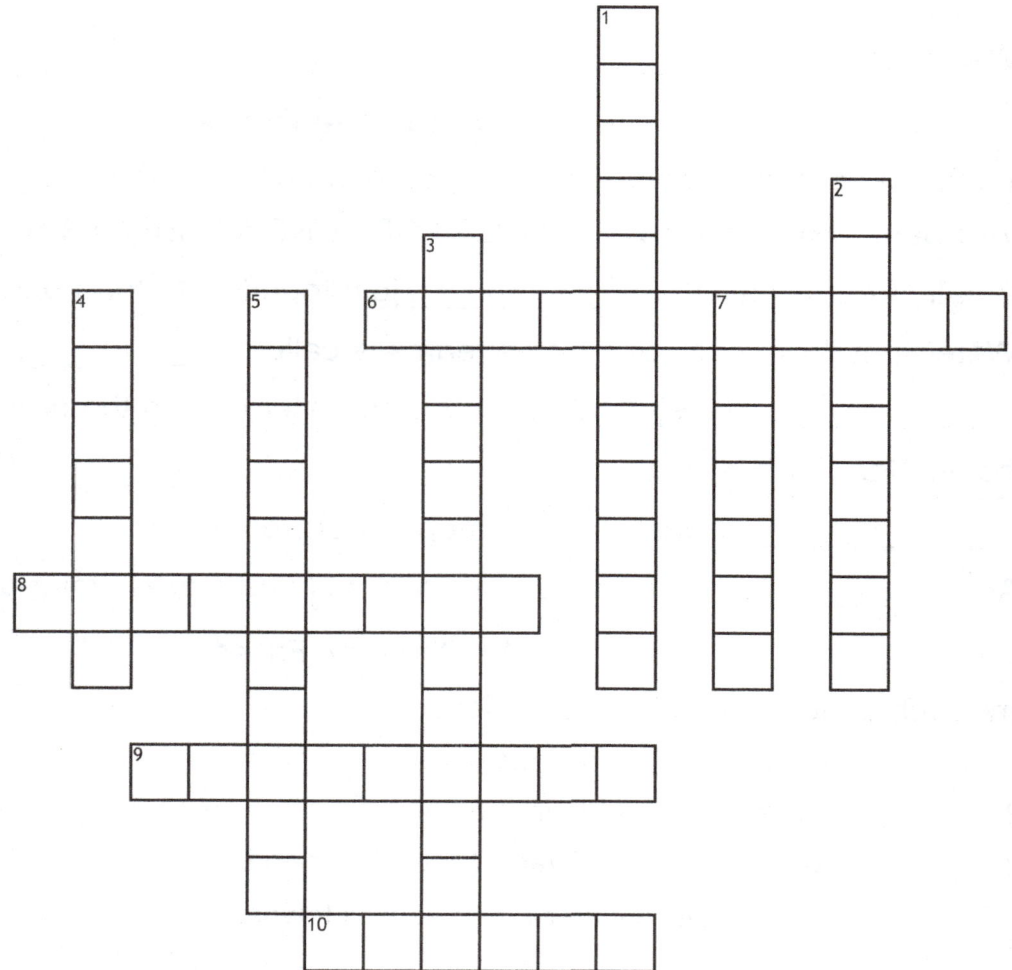

Water Systems

```
I H L T Z J E H P M G A C A X A T
U T R J G G W V A Q U I F E R I J
B H G R R L T Q A J Z Q P W F G X
P R H R R O K P W P U V G N J H X
B C X C O O H G Q A O V I T X K M
H T R I B U T A R Y T R M X I T G
A S R X U R N N U R F E A L F J C
X O N Z Q P C D R Y D P R T P B J
R U N O F F J V W K S J C S I Q I
Y W F G Q I Z D L A R F W C H O P
Y I M Y Q Q O E Y Z T E F E H E N
W B C G Q P B A N O Q E M A F V D
S E T W Q P N J I M S D R V B T R
F P R E C I P I T A T I O N R O R
D X K C Y C O N D E N S A T I O N
U M E W E T L A N D E Z N N L A X
A D A U L T B R E S E R V O I R G
```

WETLAND   WATERSHED   TRIBUTARY

RUNOFF   RESERVOIR   PRECIPITATION

GROUNDWATER EVAPORATION   CONDENSATION

AQUIFER

Sustainable design is about creating things, like buildings and products, in a way that uses less energy and resources and helps protect the planet, so it can stay healthy for future generations!

Imagine designing a better future for our planet! **SUSTAINABLE** design helps us create things that don't harm Earth. One important way is using **RENEWABLE** resources like **SOLAR** energy from the sun and **WINDPOWER** from moving air. These clean energy sources work with amazing **EFFICIENCY** to power our homes and schools.

Smart designers think about what happens to products when we're done using them. **BIODEGRADABLE** materials naturally break down instead of filling landfills. We can **RECYCLE** materials like plastic and paper to make new things, or **REPURPOSE** old items in creative ways. Some clever designers even **UPCYCLE** trash into treasure, making beautiful new products from things that would have been thrown away!

In our gardens, we can create **COMPOST** from food scraps and leaves, turning waste into rich soil for growing plants. When we design with Earth in mind, we create a cleaner, healthier world for everyone!

WORD SCRAMBLE

#	Scramble	
1	ARSOL	
2	BNISALASUET	
3	RSEEPUROP	
4	ENLBWERAE	
5	ECUCYPL	
6	EYECCLR	
7	CCIEFFNYEI	
8	OTMSCPO	
9	OWNEWIDRP	
10	OGEEAABDRDLIB	

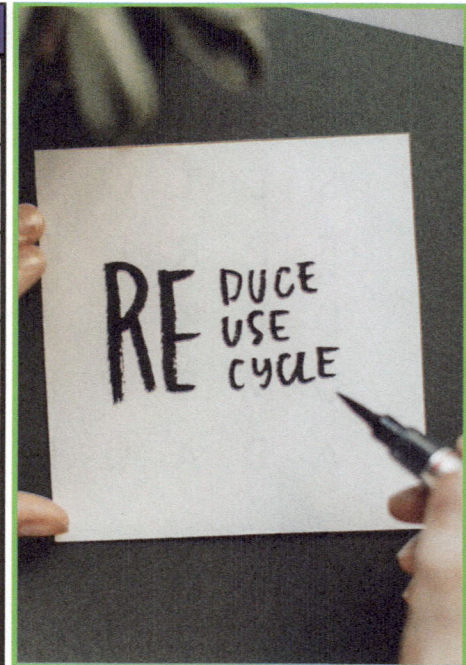

Sustainable design encourages using biodegradable materials, recycling items, repurposing things we no longer need, and upcycling old products into something new, all to help reduce waste and protect our environment!

I. Key Definitions

Write the definition for each term:

1. Sustainable: _____

2. Renewable Resources: _____

3. Biodegradable: _____

II. Fill in the Blank

Use the words from the word bank to complete each sentence.

Word Bank: SOLAR | WINDPOWER | RECYCLE | UPCYCLE | COMPOST

1. _____ energy comes from the sun.

2. _____ is energy that comes from moving air.

3. We can _____ materials like plastic and paper to make new things.

4. People can _____ trash into treasure by making new products.

5. We can make _____ from food scraps and leaves to create rich soil.

III. True or False

Mark each statement as True or False:

1. _____ Sustainable design helps protect Earth.

2. _____ Renewable resources are bad for the environment.

3. _____ We can repurpose old items in creative ways.

4. _____ Compost helps grow plants.

IV. Matching

Match the term with its description:

1. ___ Solar   A. Makes old trash into new items

2. ___ Upcycle   B. Materials that break down naturally

3. ___ Biodegradable   C. Energy from the sun

4. ___ Compost   D. Made from food scraps and leaves

V. Reflection

Why is it important to think about sustainable design in our daily lives?

_____

_____

_____

Sustainable Design

Across

4. Decayed organic material used to enrich soil

5. To transform waste into something of better quality

8. A resource that can be naturally replaced

9. Process of converting waste into reusable material

10. Energy generated by moving air

Down

1. Material that can naturally break down over time

2. Using resources without waste

3. Able to maintain without harming the environment

6. Energy that comes from the sun

7. Give an item a new use instead of throwing it away

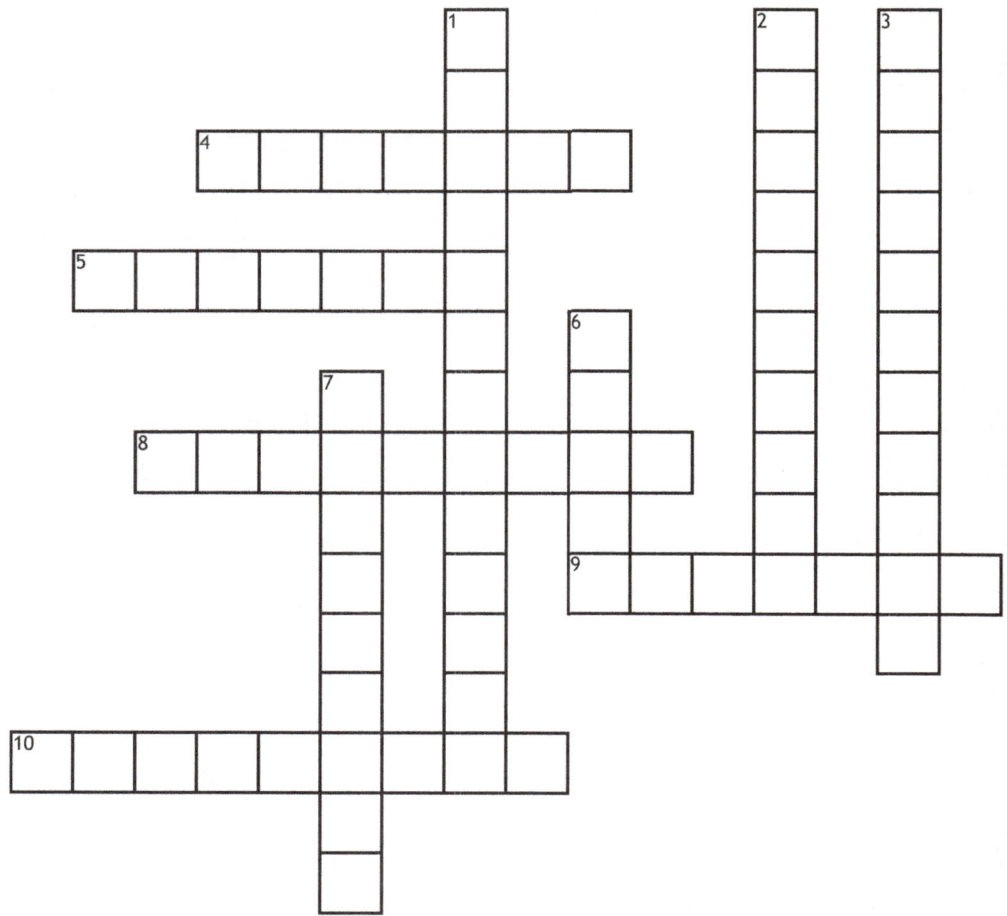

Sustainable Design

```
H B Y Y S N M L W D A N P C F U C
R Q Z N I W M H V B J M G E S Q K
E D O J N S U S T A I N A B L E B
C A B D E W R V R R I I S T R S I
Y Q K O C R I D Q X R J T P A U O
C O D O E H E N G N C A X L R H D
L H S O L A R F D P I J H N J Z E
E R P R R Z L F F P B W V U E S G
X E B J K E V D R I O K O L I K R
C N U S C B P P T Y C W L V G W A
Q E J L G X T U S G T I E Q K K D
U W C P M Y O Y R P D G E R H Y A
V A V H C V T E F P B M L N E O B
H B C O M P O S T S O D D F C R L
R L I K G Z I L K A L S S V K Y E
N E P Y V F B P K G Z S E J V A F
N B J I A V U P C Y C L E Z A C R
```

WINDPOWER   UPCYCLE   SUSTAINABLE

SOLAR   REPURPOSE   RECYCLE

RENEWABLE   EFFICIENCY   COMPOST

BIODEGRADABLE

Urban planning is the process of designing and organizing cities and towns to make them better places to live, work, and play by creating parks, roads, homes, and schools that meet the needs of the community!

Have you ever wondered how cities are planned? Urban planners work hard to create spaces where **COMMUNITY** members can live, work, and play together. They design wide **BOULEVARDS** and streets to handle **TRAFFIC**, making sure **PEDESTRIANS** have safe sidewalks too.

Cities need strong **INFRASTRUCTURE** - the systems that provide water, electricity, and other services we use every day. **TRANSPORTATION** is super important, so planners create bus routes, bike lanes, and train systems to help people get around.

ZONING rules help organize cities by deciding which areas should be **RESIDENTIAL** for homes, which should have stores, and which should have factories. Many people choose to live in a **SUBURB** just outside the busy city center.

Planners also make sure to include **GREENSPACE** like parks and playgrounds where people can enjoy nature and have fun. These spaces help make cities healthier and happier places to live!

WORD SCRAMBLE

1	ITFUNUACRETSRR	
2	NMYCOTIUM	
3	UABOVLDER	
4	ICRTAFF	
5	ATRIATTPONSONR	
6	PCAEGSENER	
7	NOINZG	
8	DPTIENESAR	
9	LANTIDRESIE	
10	UBBRSU	

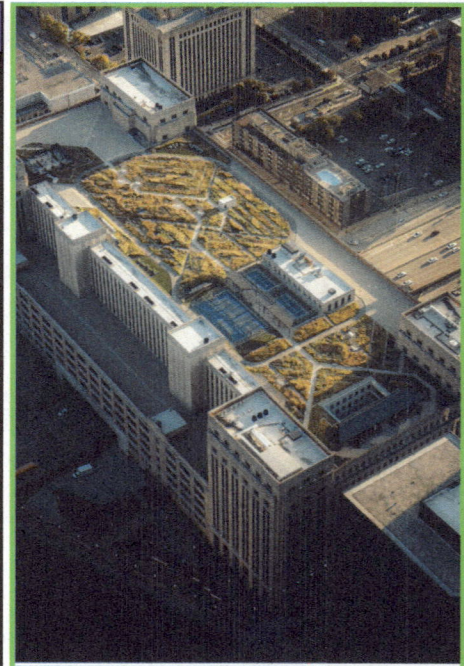

Zoning is a way for cities to divide land into different areas for specific uses, like homes, schools, parks, and businesses, ensuring that everything fits well together and makes the community a nice place to live!

I. Key Definitions

Write the definition for each term:

1. Infrastructure: _____

2. Zoning: _____

3. Greenspace: _____

II. Fill in the Blank

Use the words from the word bank to complete each sentence.

Word Bank: COMMUNITY | TRAFFIC | PEDESTRIANS | RESIDENTIAL | SUBURB

1. Urban planners create spaces where _____ members can live, work, and play.

2. Streets are designed to handle _____ and keep people safe.

3. _____ need safe sidewalks to walk on.

4. _____ areas are where people build their homes.

5. Many people live in a _____ just outside the busy city center.

III. True or False

Mark each statement as True or False:

1. _____ Cities don't need parks or playgrounds.

2. _____ Transportation helps people get around the city.

3. _____ Urban planners help design cities.

4. _____ Infrastructure provides important services we use daily.

IV. Matching

Match the term with its description:

1. ____ Infrastructure   A. Wide city streets

2. ____ Boulevards   B. Parks and playgrounds

3. ____ Greenspace   C. Systems that provide water and electricity

4. ____ Transportation   D. Buses, bikes, and trains

V. Reflection

Why do you think it's important to have good planning in a city?

_____

_____

_____

Urban Planning

Across

5. Basic systems that keep a city running

7. A community located outside the main city

8. Rules about how land can be used in different areas

10. Areas where people live in homes

Down

1. Group of people living in the same area

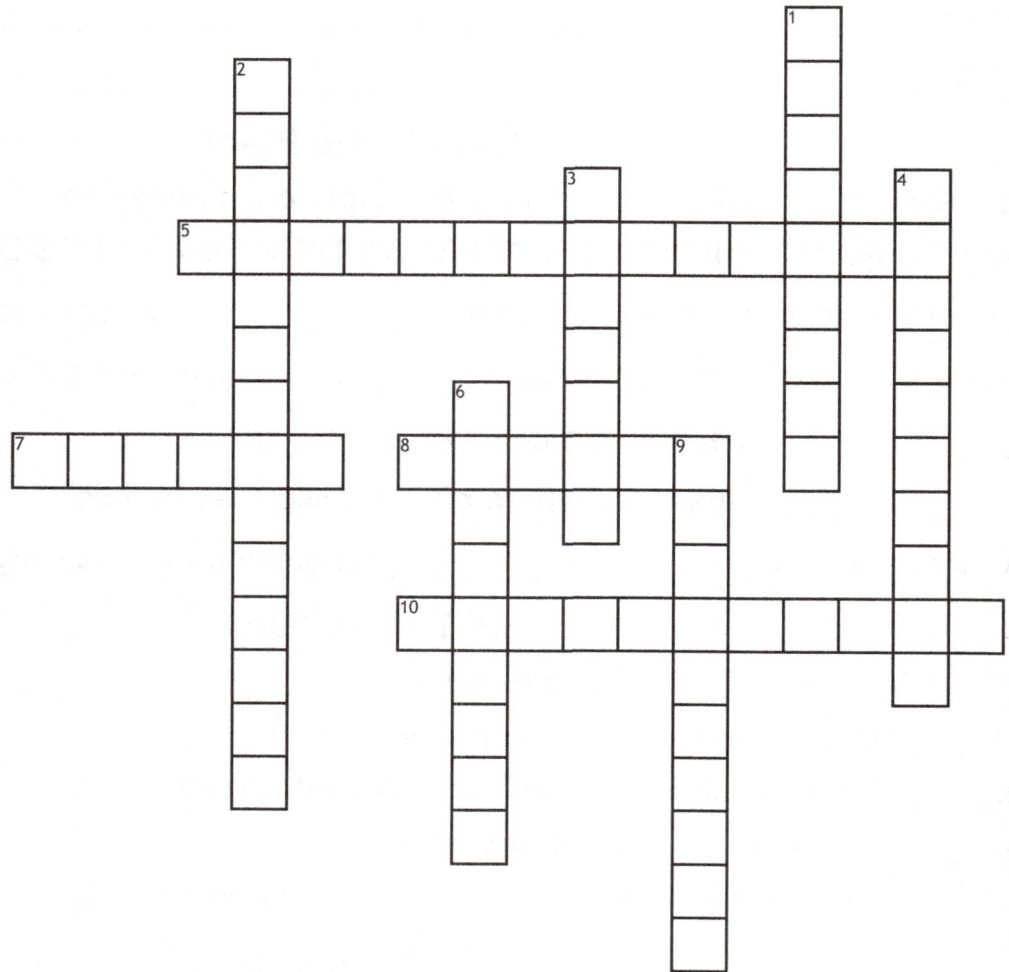

2. Ways of moving people and things around

3. The movement of vehicles on roads

4. Person walking on streets or sidewalks

6. Wide city street, often lined with trees

9. Parks and natural areas within a city

Urban Planning

```
R Y Y S U B U R B C C O H K L C N
D B A S M T R H K X U Y Y V A O W
T T T I M H W E T H L Z Q F S M T
D V R Z N B G E S W N O F L H M T
Y F I A O F O R V I T J K R H U D
J P E R N N R U E B D Q N Z V N K
R U N H E S I A L E E E B O S I X
Q I B C V A P N S E N K N V U T J
U M A Y U L W O G T V S L T A Y W
J I S Z O L S P R P R A P F I G E
C Z D Y U F K N M T P U R A W A P
P E D E S T R I A N A J C D C M L
B O P X P Y N L D Y I T G T Z E X
P V P I H Z U M P E G B I L U T S
A R F I Q L T A E H K R V O U R L
G M O H P Y Z Y J W Y N I S N Q E
S C G J A N X H T R A F F I C K O
```

ZONING   TRANSPORTATION   TRAFFIC

SUBURB   RESIDENTIAL   PEDESTRIAN

INFRASTRUCTURE   GREENSPACE   COMMUNITY

BOULEVARD

Environmental solutions are ideas and actions we take to protect our planet, like recycling, using renewable energy, and planting trees, which help keep our air clean and support wildlife!

Our planet needs heroes who can solve environmental problems! When we practice **CONSERVATION**, we protect Earth's resources for the future. Sometimes this means saving **ENDANGERED** animals by protecting their **HABITAT** from damage or destruction.

A healthy **ECOSYSTEM** includes many different **NATIVE** plants and animals living together. But sometimes **INVASIVE** species arrive and cause problems for local **WILDLIFE**. Scientists work hard to control these unwanted visitors and protect natural areas through **PRESERVATION** efforts.

POLLUTION in our air, water, and soil can harm living things, but there are solutions! **RESTORATION** projects help fix damaged environments. People clean up trash from rivers, plant trees in empty lots, and build gardens for butterflies and birds.

When we work together to solve environmental problems, we make Earth a better home for all living things. Every person can be an environmental hero by making choices that help protect nature!

WORD SCRAMBLE

1	NEAIISVV	
2	SNVAIEOCOTNR	
3	SOMEYSCTE	
4	NOUOPLTIL	
5	ITEANV	
6	DNEGNERDAE	
7	ENTTSOIRROA	
8	AATHTBI	
9	LEIFDILW	
10	OEVAEIRRNTSP	

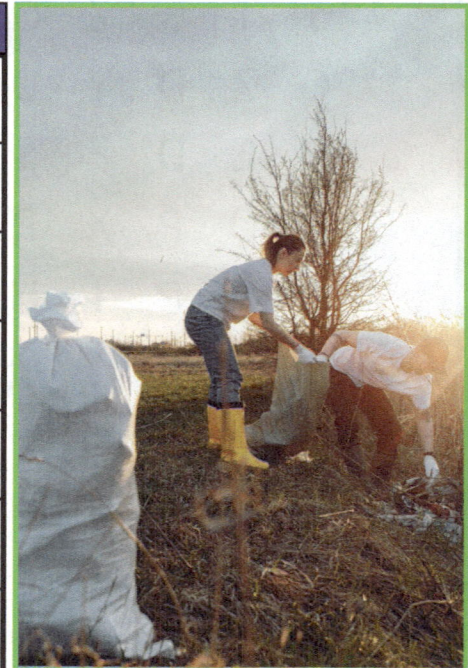

Pollution is when harmful substances make our air, water, or land dirty, but restoration means fixing those damaged places by cleaning them up and bringing back plants and animals to create a healthier environment!

I. Key Definitions

Write the definition for each term:

1. Conservation: _____

2. Ecosystem: _____

3. Habitat: _____

II. Fill in the Blank

Use the words from the word bank to complete each sentence.

Word Bank: ENDANGERED | NATIVE | INVASIVE | WILDLIFE | RESTORATION

1. _____ animals need protection to survive.

2. _____ plants and animals naturally live together in an area.

3. _____ species can cause problems for local plants and animals.

4. Local _____ can be harmed by pollution.

5. _____ projects help fix damaged environments.

III. True or False

Mark each statement as True or False:

1. _____ Only scientists can help protect the environment.

2. _____ Pollution can harm living things.

3. _____ People can plant trees to help the environment.

4. _____ Every person can be an environmental hero.

IV. Matching

Match the term with its description:

1. ___ Conservation   A. Fixing damaged environments

2. ___ Preservation   B. Where animals live

3. ___ Habitat   C. Protecting Earth's resources

4. ___ Restoration   D. Protecting natural areas

V. Reflection

How can you be an environmental hero in your daily life?

_____

_____

_____

Environmental Solutions

Across

6. A community of living things and their environment

8. Animals living in nature

9. Natural home of a plant or animal

10. The returning damaged environments to natural state

Down

1. A non-native species that harm local environments

2. Careful use and protection of natural resources

3. WW. Protecting nature from harm or change

4. Plants or animals that naturally belong in an area

5. Harmful substances in the environment

7. Species at risk of becoming extinct

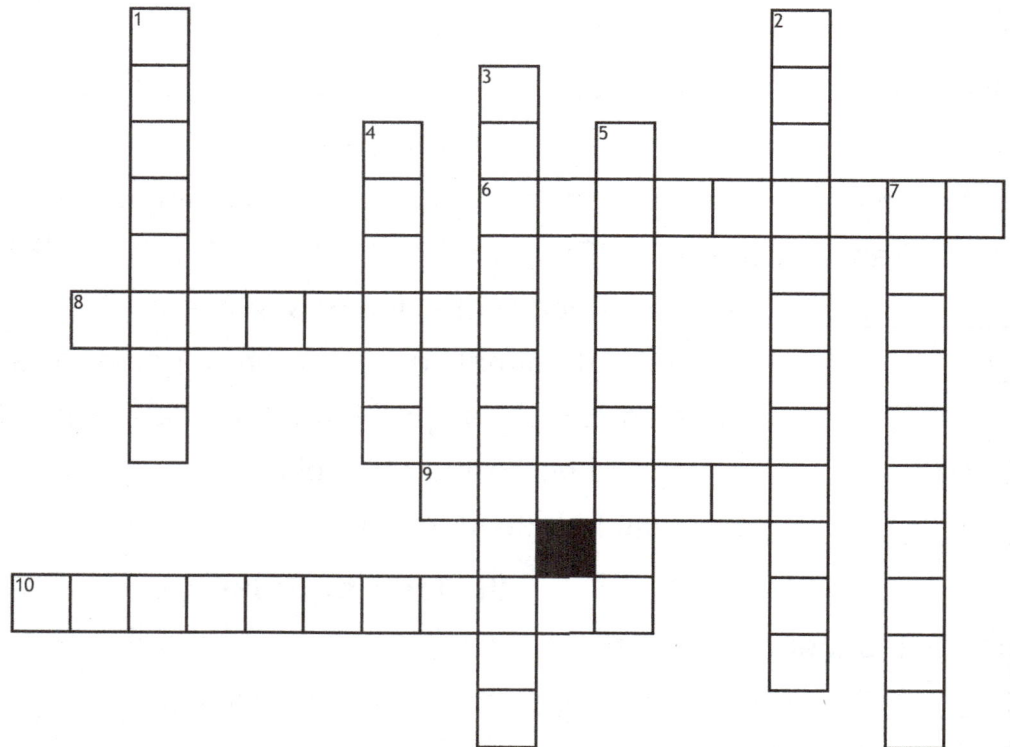

Environmental Solutions

```
T P H M N L J F E T T Q S T N X P
W I R U V J X Q A O D Q G E N C R
D X G Y T N A T I V E O M L C F E
N E O Q Q J W P H N D K M L O E S
X E N D A N G E R E D F L V N G E
E P O L L U T I O N N F C K S G R
H C J B D I P F K N S W C R E H V
S A O W J Y N K S K X D F E R P A
Y W B S I E Y V K C J N H S V U T
J I E I Y N K J A Q L M O T A B I
Q L K O T S H Z Z S Q T C O T Q O
R D K J M A T B X M I Z U R I R N
Q L M T Z T T E R H V V Z A O E A
X I H T Y N E A M H I U E T N W Q
I F L V O K K I R H B P Q I V I Q
V E A Z M B H Q B U L E Z O X G J
M M Q H T Q E U J F C Z X N Z Q V
```

WILDLIFE   RESTORATION   PRESERVATION
POLLUTION   NATIVE   INVASIVE
HABITAT   ENDANGERED   ECOSYSTEM
CONSERVATION

Waste management is the process of collecting, recycling, and disposing of trash properly to keep our communities clean and protect the environment from pollution!

Did you know we can be waste warriors? The first step is to **REDUCE** how much trash we make. When we **REUSE** items like water bottles and lunch containers, we create less garbage that needs **DISPOSAL** in a **LANDFILL**.

RECYCLING is another super solution! **SORTING** our trash helps turn old materials into new things. Paper, plastic, glass, and metal can all be recycled instead of becoming **LITTER** that harms our environment. Some materials naturally **DECOMPOSE**, like banana peels and leaves. But **HAZARDOUS** materials like chemicals and batteries need special handling because they can harm people and nature. Sometimes waste management facilities **INCINERATE** certain types of trash that can't be recycled or composted.

Being a waste warrior means thinking carefully about our trash. Every time we reduce, reuse, and recycle, we help keep Earth clean and healthy for everyone!

WORD SCRAMBLE

1	DERUCE	
2	CMEPSEDOO	
3	INGTOSR	
4	TNNECIREIA	
5	DLLINLAF	
6	NCICGEYLR	
7	DILAOPSS	
8	ZSUAHODRA	
9	UEESR	
10	TLERTI	

Recycling is when we take old items like paper, plastic, and glass and turn them into new products instead of throwing them away, helping to save resources and reduce waste!

I. Key Definitions

Write the definition for each term:

1. Reduce: _____

2. Landfill: _____

3. Hazardous Materials: _____

II. Fill in the Blank

Use the words from the word bank to complete each sentence.

Word Bank: REUSE | RECYCLING | SORTING | LITTER | DECOMPOSE

1. _____ items like water bottles helps create less garbage.

2. _____ helps turn old materials into new things.

3. _____ our trash helps us separate different materials.

4. _____ can harm our environment.

5. Some materials naturally _____, like banana peels and leaves.

III. True or False

Mark each statement as True or False:

1. _____ We should throw everything in the trash.

2. _____ Paper, plastic, glass, and metal can be recycled.

3. _____ Batteries need special handling.

4. _____ Being a waste warrior helps keep Earth clean.

IV. Matching

Match the term with its description:

1. ___ Reduce   A. Using items again

2. ___ Reuse   B. Making less trash

3. ___ Recycling   C. Breaking down naturally

4. ___ Decompose   D. Turning old materials into new things

V. Reflection

What can you do to be a waste warrior at home or school?

_____

_____

_____

Waste Management

Across

2. This is converting waste into new materials

4. Process of getting rid of waste

6. This is dangerous to health or environment

7. This is separating different types of waste

9. Trash left in wrong places

10. Use less of something

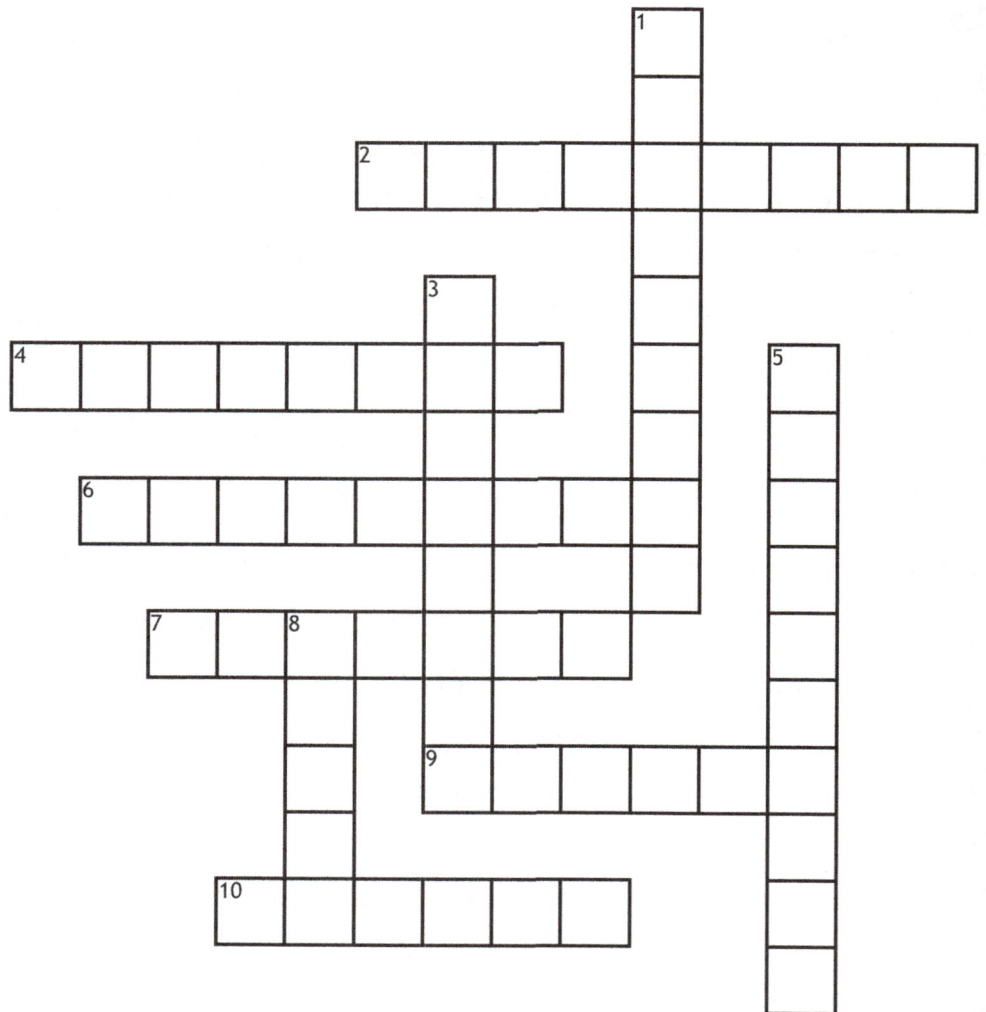

Down

1. To break down naturally over time

3. Place where trash is buried in the ground

5. To burn waste completely

8. Use something again

Waste Management

```
U H F S G H D E C O M P O S E G R
Q R A R J U A V S L W J H B F C M
T C R A E F P Z N D D G H W E F A
W A R E L I D S A N Y I T R J F T
H L E I C J A O V R V B N T N R K
M B D N N Y O R V S D I H R T H Y
S E U J S C B T D H O O U E Q F C
M O C Q C Z I I E I E A U U S X C
V O E S D G I N R K S F N S I V Q
K W L L K A S G E J R P V E E F Q
Z C I A R Z V W C R K H O K J V A
Q Z T N N F F C Y M A U N S T R Q
N D T D Q J M P C X F T W V A F B
M W E F A D L I L W S Y E O I L Q
C B R I G J J K I I P E G L H X N
Q A X L Q Y J G N F T J F A J J R
P Z K L B N D L G E R S I Y X M F
```

SORTING   REUSE   REDUCE

RECYCLING   LITTER   LANDFILL

INCINERATE   HAZARDOUS   DISPOSAL

DECOMPOSE

Robotics engineering is the field where we design and build robots that can help us with tasks, explore new places, and even make our lives easier and more fun!

Imagine being an **ENGINEER** who creates amazing machines! Robotics engineers design **ROBOTS** that can do incredible things. First, they make a **PROTOTYPE** to test their ideas. They carefully plan how to connect each **CIRCUIT** and **MOTOR** to make their robot move.

Every robot needs a **BATTERY** for power and **SENSORS** to understand the world around it. These **SENSORS** help robots detect things like light, sound, or movement. But a robot can't do anything without proper **CODING**. When engineers do **PROGRAMMING**, they're giving the robot specific instructions to follow.

One of the most exciting things about robotics is **AUTOMATION** - this means robots can do jobs all by themselves! From helping to build cars in factories to exploring other planets, robots make our lives easier and more interesting. With robotics engineering, we can create machines that help solve problems and make the world a better place.

WORD SCRAMBLE

1	CCITUIR	
2	NRIEENGE	
3	TMOOR	
4	ORTOB	
5	RAETBYT	
6	RSOESN	
7	RRMGPOA	
8	PTOOYTERP	
9	NTIUAAOTMO	
10	DIGCON	

Automation is using technology and machines to do tasks automatically without needing a person to do them, which helps save time and makes work easier!

I. Key Definitions

Write the definitions for these important terms:

1. Prototype: _____

2. Sensors: _____

3. Automation: _____

II. Fill in the Blank

Word Bank: BATTERY   PROGRAMMING   CIRCUIT   MOTOR   ROBOTS

1. Robotics engineers design _____ that can do incredible things.

2. Engineers need to connect each _____ and _____ to make their robot move.

3. Every robot needs a _____ for power.

4. When engineers do _____, they're giving the robot specific instructions to follow.

III. True or False

Write T for True or F for False:

1. _____ Robots can help build cars in factories.

2. _____ Robots never need sensors to understand their environment.

3. _____ Robots can explore other planets.

4. _____ Robots can work without any power source.

IV. Matching

Match the term with its correct description by writing the letter in the blank:

1. _____ Sensors   A. Gives robots power to work

2. _____ Battery   B. Detect things like light and sound

3. _____ Programming   C. Robots working by themselves

4. _____ Automation   D. Giving robots instructions to follow

V. Reflection

How do you think robots help make our world a better place?

_____

_____

_____

Robotics Engineering

Across

3. A device that detects changes in the environment

6. Path that electricity follows in a robot

7. First version of a robot design

8. The instructions that tell a robot what to do

9. Person who designs and builds robots

10. Writing instructions for computers and robots

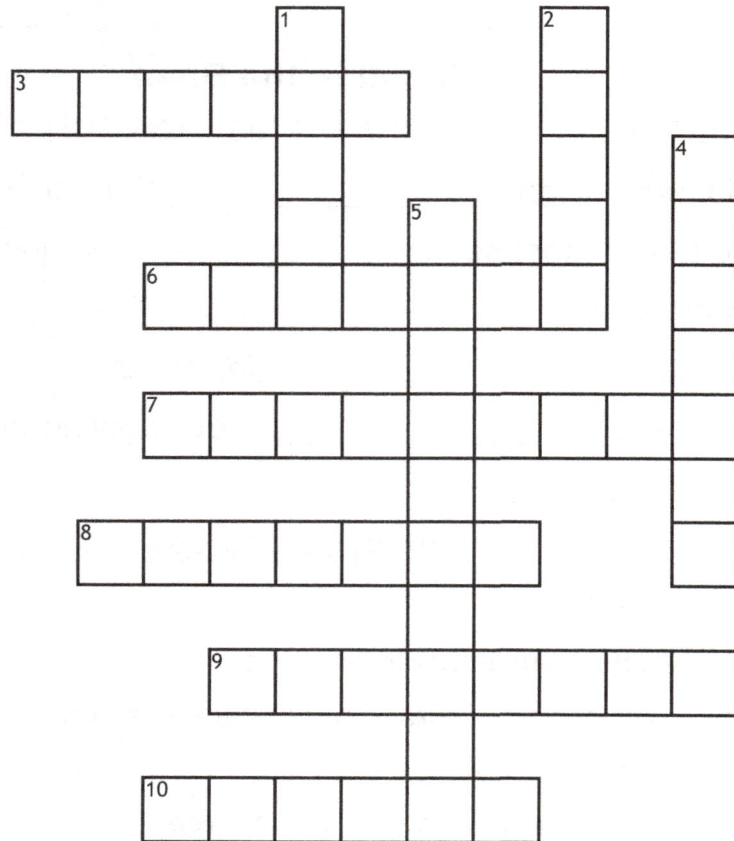

Down

1. A part that creates movement in a machine

2. A machine that can be programmed to do tasks

4. Power source that makes robots run

5. When machines do work without human help

Robotics Engineering

```
P S L E S U A O N S L E X I P L H
R Q A Y N B A T T E R Y W G B R F
O B R P G G B C J R U X A Q G P V
T O N Z I P I P O T P A H U A G O
O B J U U J P N C D N X K R K L P
T I M R V H A P E C I R C U I T N
Y G C Y I H V E O E S N F C D N M
P H I Q M O T O R U R L G L R X R
E R Y D A U T O M A T I O N L H R
K A O I X R A F Z E N Y U J J J U
E R P G C R B F X Q T S D Q X M E
R H I P R V K T N I Q E B M G S V
R Y V O E A X B H Q V N B F T S V
O W U B M S M O I E R S F U Z F X
O M E K E B W I V R T O X G V P M
X Y Y A P X R O B O T R W Q F N T
Z A Z D A I L R K T H W H B D Z Q
```

AUTOMATION   PROTOTYPE   CODING
ROBOT   CIRCUIT   ENGINEER
BATTERY   MOTOR   PROGRAM
SENSOR

Electrical circuits are paths that allow electricity to flow through, powering things like lights and toys when all the parts are connected properly!

Have you ever wondered how electricity travels through wires? A **CIRCUIT** is like a road that **ELECTRONS** travel on. These tiny particles need a **BATTERY** to give them the energy to move. The **VOLTAGE** from the battery is like pushing these electrons forward, creating an electrical **CURRENT**.

Some materials are excellent **CONDUCTORS**, like copper wire, which lets electricity flow easily. Other materials are **INSULATORS**, like rubber or plastic, which keep the electricity safely inside the wires. When building circuits, we often add **RESISTANCE** to control how fast the electricity flows, just like speed bumps on a road.

A **SWITCH** is an important part that lets us control when electricity flows. Scientists can build circuits in different ways - one cool way is a **PARALLEL** circuit, which gives electricity multiple paths to travel. This is like having several roads leading to the same place! Understanding circuits helps us power everything from tiny toys to huge buildings.

WORD SCRAMBLE

#	Scramble	
1	LCONEREST	
2	TLAVGOE	
3	RSIOUTLNA	
4	CCIURIT	
5	SECRSTNAEI	
6	APLLALER	
7	DURCTNCOO	
8	REUNRTC	
9	RTBEAYT	
10	WICTSH	

Conductors are materials that allow electricity to flow easily, like copper wires, while insulators are materials that block electricity, like rubber, and resistance is the measure of how much a material opposes the flow of electricity!

I. Key Definitions

Write the definitions for these important terms:

1. Circuit: _____

2. Conductors: _____

3. Insulators: _____

II. Fill in the Blank

Word Bank: ELECTRONS   BATTERY   SWITCH   VOLTAGE   CURRENT

1. A _____ gives energy to electrons to help them move.

2. _____ are tiny particles that travel through circuits.

3. _____ pushes electrons forward.

4. When electrons move, they create electrical _____.

5. A _____ controls when electricity flows.

III. True or False

Write T for True or False:

1. _____ Copper wire is a good conductor of electricity.

2. _____ Rubber lets electricity flow easily through it.

3. _____ Parallel circuits give electricity multiple paths to travel.

4. _____ Circuits help power both small toys and large buildings.

IV. Matching

Match the term with its correct description by writing the letter in the blank:

1. _____ Battery   A. Controls when electricity flows

2. _____ Switch   B. Keeps electricity inside wires

3. _____ Insulator   C. Gives energy to electrons

4. _____ Resistance   D. Controls how fast electricity flows

V. Reflection

How is a circuit like a road? Use information from the article in your answer.

_____

_____

_____

Electrical Circuits

Across

2. Tiny particles that carry electricity

5. The electrical pressure that pushes current

6. Material that blocks electrical flow

9. Device that turns a circuit on and off

10. A complete path that electricity can flow through

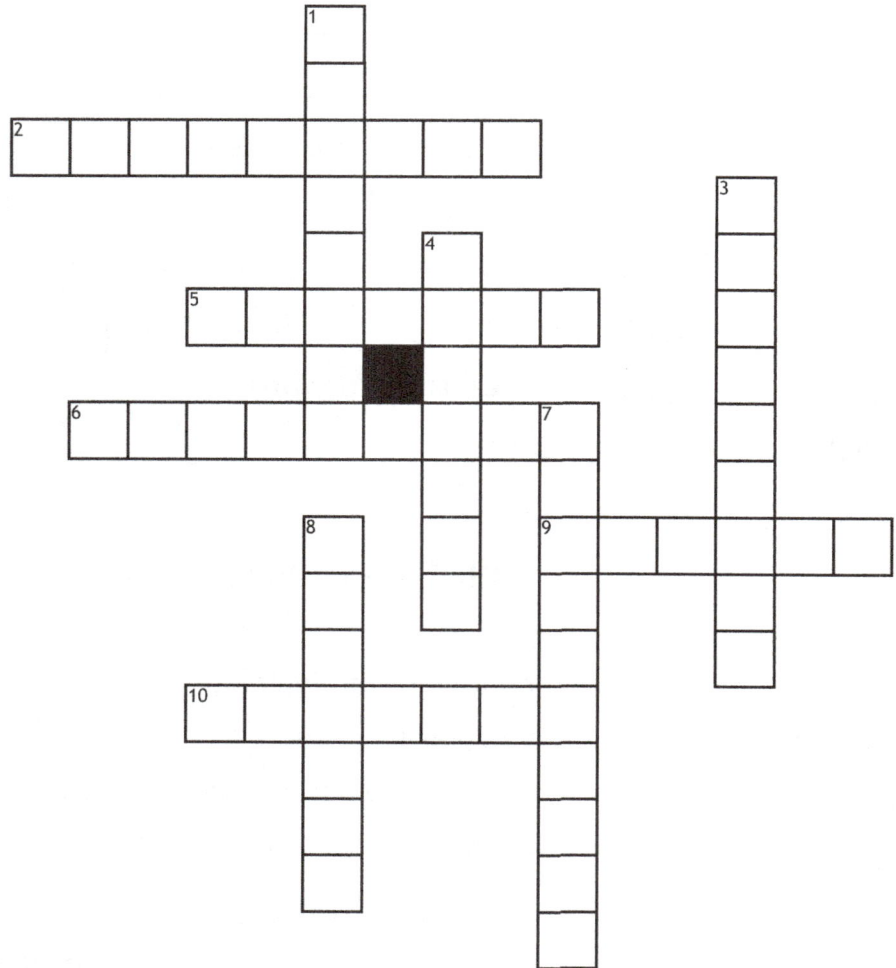

Down

1. Circuit type where electricity has multiple paths

3. Material that lets electricity flow easily

4. Power source that provides electricity

7. Force that slows down electrical flow

8. The flow of electricity through a wire

Electrical Circuits

```
X O X H G N W H F Y W J T K H W D
N B A T T E R Y Y O O F K K J R H
O Q T S D E E C I R C U I T K V D
Q O S G K E I M H N V J G S R G S
P O Y M U A K C E Z S E T G E B E
U P A R A L L E L J X U G P S A H
A C Y C B F C L D Y N B L I I H R
W K U N L K Y E N K C E D A S C C
W U I R U V M P L W J Q K O T Z P
Z V Y Q R X B X O E V X E P A O S
U V T W Q E L A S Z C M N P N T R
M O V U W J N P W T J T R Y C L P
X L Q K D J P T I N Z X R G E T I
B T G L R H N O T O M H L O W M J
K A C I W R G E C E I G F S N O N
S G W V L V C A H J D E S G C S K
C E C Y C O N D U C T O R T D C D
```

PARALLEL   ELECTRONS   RESISTANCE

SWITCH   INSULATOR   CONDUCTOR

BATTERY   VOLTAGE   CURRENT

CIRCUIT

AVIATION is an exciting field where engineers design amazing flying machines! The science of flight depends on four main forces. **THRUST** pushes the aircraft forward, often created by a spinning **PROPELLER** or **TURBINE** engine. **LIFT** pulls the plane upward into the sky, while **DRAG** tries to slow it down.

Aeronautical engineering is the branch of engineering that focuses on designing and building airplanes and other flying machines to help them soar through the sky safely and efficiently!

The **FUSELAGE** is carefully designed to be **AERODYNAMIC**, letting it slice through the air like a bird. Pilots control their **ALTITUDE** using various parts of the aircraft. Small moving surfaces called **AILERONS** help the plane turn and roll in the sky.

Modern aircraft can fly higher, faster, and more safely than ever before. Whether it's a small propeller plane or a massive passenger jet, each aircraft is a marvel of engineering. The next time you see an airplane soaring overhead, remember all the careful planning and science that helps it fly through the clouds!

WORD SCRAMBLE

1	AOTNAIVI	
2	TIUTDELA	
3	IFLT	
4	DGAR	
5	UHSTRT	
6	GLSFAEEU	
7	NEIRBTU	
8	LPLEPORRE	
9	OARINLE	
10	IOAENMYRDAC	

Aerodynamics is the study of how air moves around things, like airplanes or cars, and helps us understand how to make them go faster and stay stable while they move!

I. Key Definitions

Write the definitions for these important terms:

1. Aviation: _____

2. Thrust: _____

3. Ailerons: _____

II. Fill in the Blank

Word Bank: LIFT   DRAG   FUSELAGE   PROPELLER   ALTITUDE

1. A _____ or turbine engine helps push aircraft forward.

2. _____ pulls the plane upward into the sky.

3. _____ tries to slow the plane down.

4. The _____ is designed to slice through the air like a bird.

5. Pilots control their _____ using various parts of the aircraft.

III. True or False

Write T for True or False:

1. _____ Modern aircraft can fly higher and faster than ever before.

2. _____ There are four main forces involved in flight.

3. _____ Only small propeller planes are marvels of engineering.

4. _____ The fuselage is designed to be aerodynamic.

IV. Matching

Match the term with its correct description by writing the letter in the blank:

1. _____ Thrust   A. Slows the plane down

2. _____ Lift   B. Helps plane turn and roll

3. _____ Drag   C. Pushes plane forward

4. _____ Ailerons   D. Pulls plane upward

V. Reflection

What makes airplanes amazing? Use information from the article in your answer.

_____

_____

_____

Aeronautical Engineering

Across

4. Force that slows aircraft down

6. Force that makes aircraft rise into the air

9. Height of an aircraft above the ground

10. Main body of an aircraft

Down

1. Shaped to move smoothly through air

2. Engine part that spins to create power

3. Spinning blades that help create thrust

5. Moving flap on wings that helps aircraft turn

7. Force that pushes aircraft forward

8. Science of flying aircraft

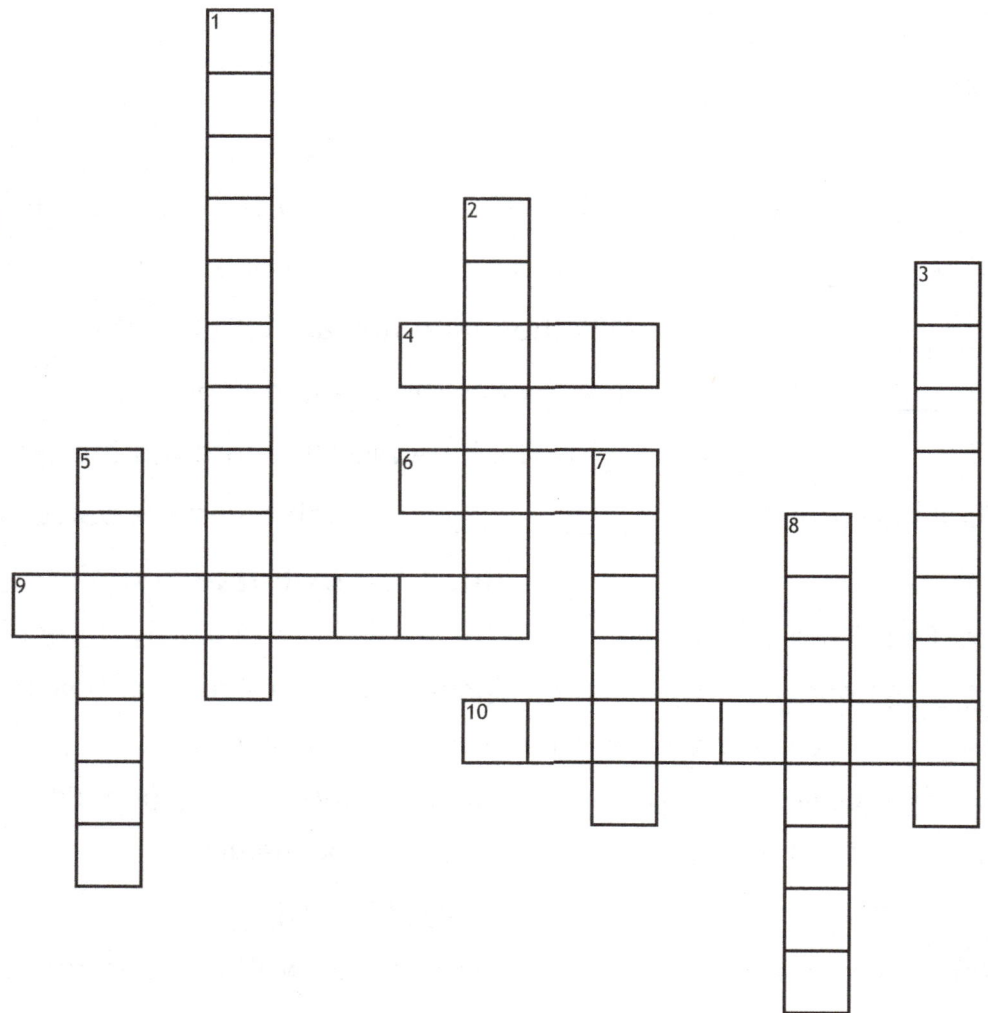

Aeronautical Engineering

```
X D O L N N X M H E X F V L T X A
O D Q S G M J B A J N V F G I D F
L M J A P F O X T G A I H T M F P
P H G Z Q U P W P U E U P K D N T
X W A B Y A N A F R R U O W V Z L
C V R M L L F H U W O B Y S L O V
V P X Z X T A J S I D P I U X Y X
B F B B O I S E E H Y B E N E F G
V I V D X T D C L L N L F L E W U
J V W H L U N I A A A C H N L S U
U E A I A D A Z G V M I K X W E I
U D R A G E L T E I I H L M U J R
H O R Y C I A H E A C R D E X K G
I S B U U K S R P T Z M U A R A B
A Y O W T C F U H I A D D F G O S
I K G X S I G S G O O H K Y L F N
S J T L Q X I T Y N C D F Z R V L
```

AERODYNAMIC   **AVIATION**   **ALTITUDE**
AILERON   **FUSELAGE**   **TURBINE**
PROPELLER   **DRAG**   **LIFT**
THRUST

BIOMEDICAL engineering helps doctors take care of patients in amazing ways! These special engineers create **DEVICES** that can see inside our bodies and help people heal. For example, they design **SCANNERS** that let doctors look at bones and organs without surgery.

When someone loses an arm or leg, engineers can create **PROSTHETIC** limbs that work almost like real ones. Some are even **ROBOTIC**, letting people move and grab things naturally! Tiny **IMPLANTS** can help hearts beat better or help deaf people hear again.

Special **SENSORS** and **MONITORS** keep track of important body signals like heartbeats and breathing. These machines help doctors make **DIAGNOSTIC** decisions about patient care. Engineers even work with **TISSUE** samples to create new ways to heal injuries.

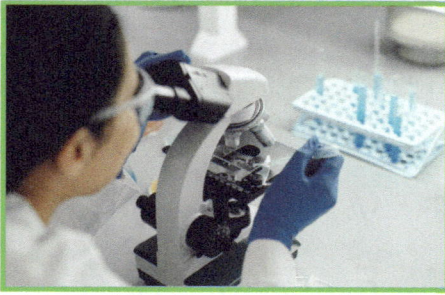

Biomedical engineering is the field that combines science and medicine to create devices and technology, like prosthetics and medical imaging machines, to help improve people's health and well-being!

Every day, biomedical engineers work hard to invent new tools that make medical care better and help people live healthier lives. It's like being a superhero who uses science to help others!

WORD SCRAMBLE		
1	OCRBIOT	
2	ISGDANCOTI	
3	NSEOSR	
4	ONIORMT	
5	ITUSES	
6	STTREHPOCI	
7	ECIVDE	
8	ACENNSR	
9	LTNPMAI	
10	ADLICMOBEI	

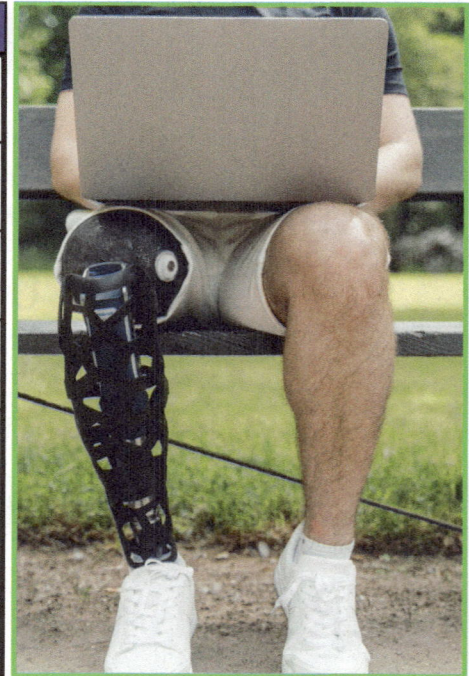

Modern prosthetics use advanced technology, like sensors and robotics, to create highly functional artificial limbs that can move more naturally and help people perform everyday tasks with greater ease!

I. Key Definitions

Write the definitions for these important terms:

1. Biomedical Engineering: _____

2. Prosthetic: _____

3. Implants: _____

II. Fill in the Blank

Word Bank: SCANNERS   SENSORS   DEVICES   TISSUE   ROBOTIC

1. Biomedical engineers create _____ that can see inside our bodies.

2. _____ let doctors look at bones and organs without surgery.

3. Some artificial limbs are _____, letting people move naturally.

4. _____ and monitors track important body signals.

5. Engineers work with _____ samples to create new ways to heal injuries.

III. True or False

Write T for True or False:

1. _____ Biomedical engineers help doctors take care of patients.

2. _____ Monitors can track heartbeats and breathing.

3. _____ Prosthetic limbs never work like real ones.

4. _____ Biomedical engineers help people live healthier lives.

IV. Matching

Match the term with its correct description by writing the letter in the blank:

1. _____ Scanners   A. Watch body signals

2. _____ Sensors   B. Help hearts beat better

3. _____ Implants   C. Look inside the body

4. _____ Devices   D. Help people heal

V. Reflection

How are biomedical engineers like superheroes? Use information from the article in your answer.

_____

_____

_____

Biomedical Engineering

Across

4. Tool that helps find health problems

5. Machine or tool made to help patients

6. Moving and working like a machine

7. Group of similar cells in the body

9. Device placed inside the body to help it work

10. Tool that measures body signals

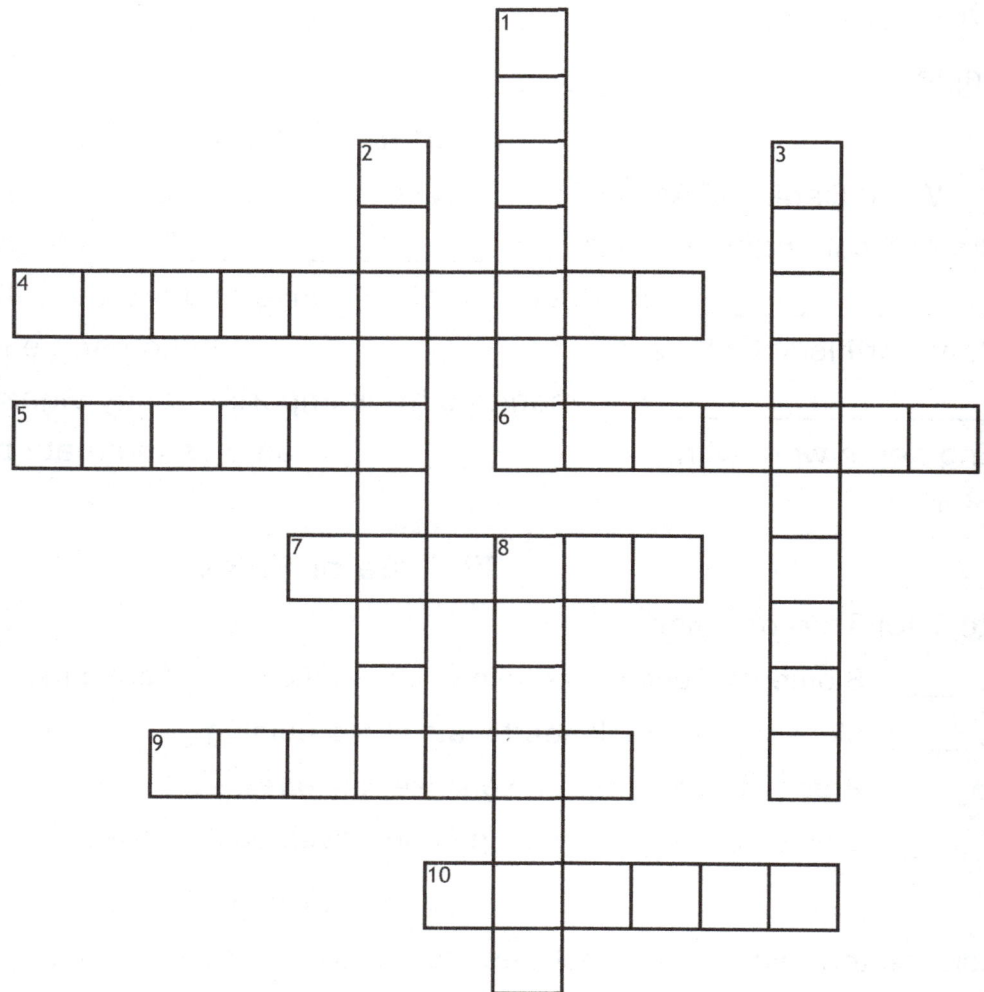

Down

1. Machine that watches body signals

2. Related to medicine and engineering

3. An artificial body part that replaces a missing one

8. Machine that takes pictures inside the body

Biomedical Engineering

```
G C W K J D V B S N A J Q Q Y C O
A K D F A Z N H P C C S N Y U R J
L R P I I W W C O L A H K B Y A G
G I D Z A D Q O H R K N H F X I J
I P U K I G L A Y J Z I N C F Y X
R X V V P P N Z Z T V M C E L W R
M Z X R G Z D O H W F R G I R G C
X F S L B H B A S L S U O G Q O F
P T A P P M G V I T E D D P L H Q
Y T I S S U E S W N I S E N S O R
P R O S T H E T I C O C H R F D D
M O N I T O R M B T D A W X H J Y
V X Y K G G B B G L A E O X M I O
U Q Q I M P L A N T W J V M K R J
F W N E H Q K W X D V G O I R U M
T A V M R S C B I O M E D I C A L
R O B O T I C X J C Y S D N Y E D
```

DIAGNOSTIC   MONITOR   ROBOTIC
DEVICE   TISSUE   SCANNER
BIOMEDICAL   SENSOR   IMPLANT
PROSTHETIC

Future engineering challenges include finding new ways to create clean energy, build smarter cities, and design safer transportation to help our planet and improve our lives!

The future needs smart engineers to solve big problems! One major challenge is making **SUSTAINABLE** cities that work better for people and the planet. Engineers are creating **RENEWABLE** energy sources like solar and wind power to replace fossil fuels.

INNOVATION is happening everywhere - from **DIGITAL** technology that makes our lives easier to **NANOTECH** that can help clean up pollution. **ARTIFICIAL** intelligence helps us solve complex problems faster than ever before.

Engineers are also finding new ways to **RECYCLE** materials and make more things **BIODEGRADABLE**. This helps protect our **ECOSYSTEM** from harmful waste. They're working hard to solve **CLIMATE** challenges by designing cleaner cars, better buildings, and smarter ways to grow food.

The future is exciting because engineers keep finding creative solutions to help people and the planet! Every new invention brings us closer to a world where technology and nature work together in harmony.

WORD SCRAMBLE

1	IAAFRLCTII	
2	ESOYSTCME	
3	TDAGILI	
4	LWANEREEB	
5	CEATNHNO	
6	ENASLTAIBUS	
7	ACEILMT	
8	GEEODBLIDBAAR	
9	RCECLEY	
10	ONTNIVAION	

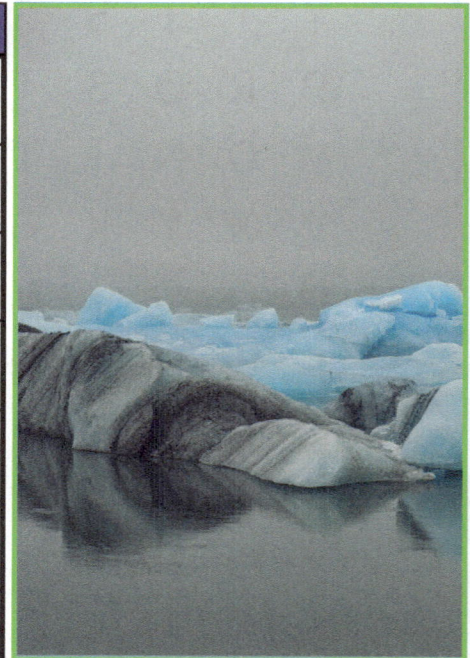

Climate challenges are problems caused by changes in the Earth's weather, like global warming and extreme storms, which affect our environment, wildlife, and how we live, making it important for us to find ways to protect our planet!

I. Key Definitions

Write the definitions for these important terms:

1. Sustainable: _____

2. Renewable Energy: _____

3. Innovation: _____

II. Fill in the Blank

Word Bank: DIGITAL   ECOSYSTEM   RECYCLE   CLIMATE   ARTIFICIAL

1. _____ technology makes our lives easier.

2. _____ intelligence helps solve complex problems quickly.

3. Engineers find new ways to _____ materials.

4. We need to protect our _____ from harmful waste.

5. Engineers are working to solve _____ challenges.

III. True or False

Write T for True or False:

1. _____ Solar and wind power are types of renewable energy.

2. _____ Engineers only work on making new cars.

3. _____ Technology and nature can work together in harmony.

4. _____ Engineers are creating ways to clean up pollution.

IV. Matching

Match the term with its correct description by writing the letter in the blank:

1. _____ Renewable Energy   A. Makes things break down naturally

2. _____ Digital   B. Uses sun and wind power

3. _____ Biodegradable   C. Makes life easier

4. _____ Nanotech   D. Helps clean pollution

V. Reflection

How are engineers helping to make the future better? Use information from the article in your answer.

_____

_____

_____

Future Engineering Challenges

Across

4. New idea or invention that solves problems

8. A community of living things and their environment

9. The long-term weather patterns of an area

10. Energy that can be naturally replaced

Down

1. A technology that works with tiny particles

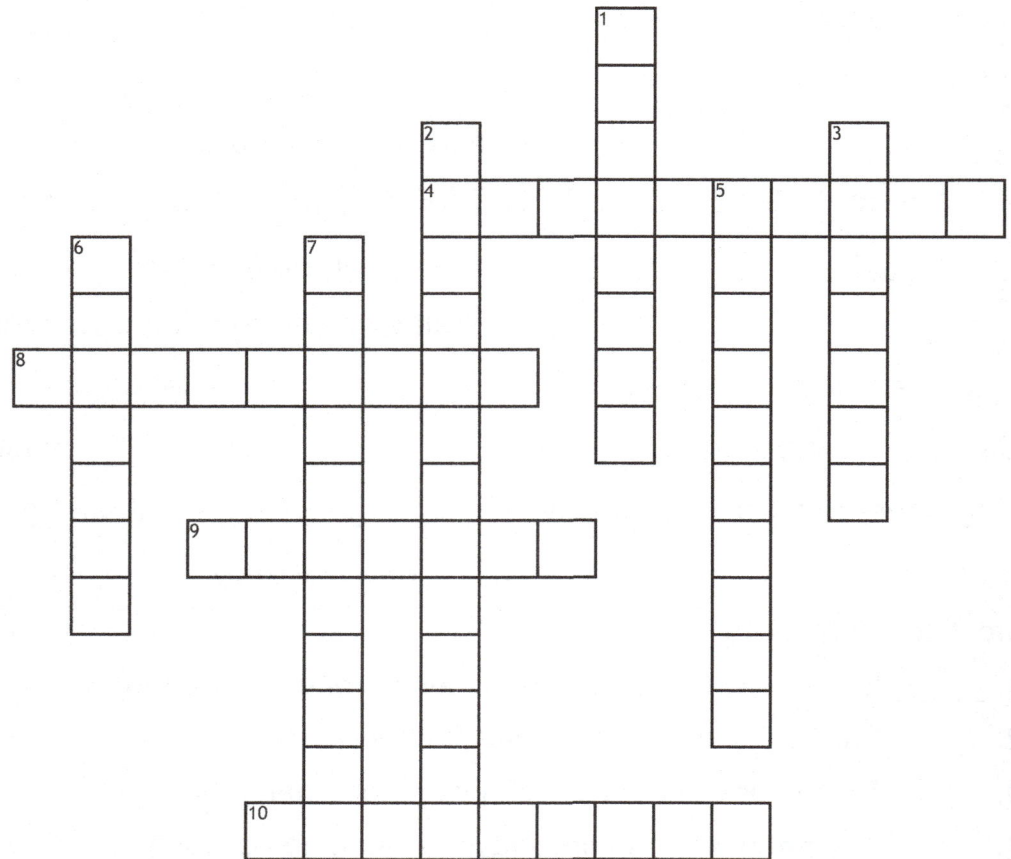

2. The materials that break down naturally

3. Using computer technology

5. Made by humans, not found in nature

6. Process of making new things from old materials

7. Using resources that won't run out

Future Engineering Challenges

```
Y  S  B  A  R  T  I  F  I  C  I  A  L  S  I  V  V
M  Z  R  I  J  V  C  A  E  R  B  Z  Z  E  W  V  R
M  A  M  H  A  J  K  L  B  C  W  N  N  L  F  E  S
V  R  E  C  Y  C  L  E  I  Y  N  G  T  T  D  I  F
N  A  N  O  T  E  C  H  O  M  X  Q  W  T  E  N  S
F  L  T  V  O  C  I  C  D  W  A  E  Z  P  J  N  U
G  U  Z  V  T  P  T  O  E  B  Z  T  T  Z  L  O  U
H  E  D  Z  I  T  A  Y  G  W  R  X  E  L  E  V  B
C  B  C  I  J  J  X  P  R  B  E  T  D  X  Y  A  H
M  G  X  O  G  V  F  K  A  V  N  S  F  C  Q  T  D
Y  J  C  S  S  I  B  P  D  U  E  X  T  I  I  I  C
L  X  U  N  A  Y  T  Q  A  Z  W  P  Z  Z  V  O  D
F  S  I  H  F  A  S  A  B  T  A  K  Z  I  G  N  B
X  J  G  Y  T  Q  Z  T  L  P  B  X  G  R  X  S  M
F  T  G  F  C  J  H  F  E  S  L  Y  T  R  C  G  C
J  L  W  B  Q  B  B  R  U  M  E  F  Y  T  P  T  Z
J  S  U  S  T  A  I  N  A  B  L  E  H  R  E  L  S
```

BIODEGRADABLE   ECOSYSTEM   DIGITAL

CLIMATE   RECYCLE   ARTIFICIAL

NANOTECH   INNOVATION   RENEWABLE

SUSTAINABLE

Your greatest superpower is your ability

to choose one thought over another.

Choose positive thoughts over negative

to live your very best life.

So much more: 3andB.com

Made in the USA
Las Vegas, NV
18 February 2025